Seal

SEAL

Fiona

Foreword by
VIRGINIA McKENNA

MAINSTREAM
PUBLISHING
EDINBURGH AND LONDON

First published in Great Britain in 1995 by
MAINSTREAM PUBLISHING COMPANY (EDINBURGH) LTD
7 Albany Street
Edinburgh EH1 3UG

ISBN 1 85158 744 6

A catalogue record for this book is available from the British Library

Typeset in Perpetua by Litho Link Ltd, Welshpool, Powys, Wales
Printed and bound in Great Britain by
Butler & Tanner Ltd, Frome and London

Contents

to the People of the Sea

Fiona

Acknowledgments

I would like to thank the following for assisting the cause of this book and the seals.

John Robins and Val Johnston and Animal Concern.

Don and Dot Bowness and the Islay and Jura Seal Action Group.

Ross Flett and Maureen Bain, Orkney Seal Rescue.

Ian Robinson and Dougie Walker, RSPCA Wildlife Hospital, Norfolk.

Virginia McKenna, The Born Free Foundation.

A special thanks to all my family for sharing their home, bath and TV with all the seals over the years.

Foreword

How many of us dream of escaping to some remote and unspoilt part of the world, far from the stress and pressures of urban life? Somewhere where horizons are truly distant, and sounds of birds, wind and water replace the clamour of traffic and machines. Perhaps there are more of us than we realise who fantasise in that way, but who never have their dream fulfilled. Trapped, maybe, in an established routine – or afraid to risk such a leap into the unknown.

No such fears or apprehensions were in Fiona Middleton's mind. The word seal (she was born in the Kent village of that name), music and her sensitivity to wild places and creatures carried her inevitably nearly 20 years ago to the Hebridean Island of Islay where she and her husband, George, were to find their spiritual home and raise their children.

This book is, in a way, unique, as it tells a story that could not possibly be told by anyone else. Communicating with seals through music – as fables tell us happened with dolphins in ancient times – Fiona wove a bond of musical empathy that deepened as the years passed. For a long time this relationship between the girl on the salt-sprayed rocks or in the bobbing boat, and her glistening 'sea-people' was a very private and personal affair. But a series of events emerged which made her realise that she could use this gift, this experience, to bring the plight of these wonderful animals to public attention.

As John Robins writes in his postscript, the message Fiona had to tell was a positive and beautiful one. Showing how extraordinary and intelligent the seals are, how they responded to her violin and to her singing, would make people realise how cruel and inhumane it was to slaughter them and to desecrate the environment which was their natural home.

What emerges so strongly in this book is the integrity of Fiona's life. Her deep love of the ocean and its inhabitants, of her island home with its ruined castle, spirits, folklore, wild flowers and animals is united, it would seem, in a completely natural way with her ability to cope with life on a very practical level.

No one living in a dream-world could manage to go out in all weathers to rescue wounded and dying seal pups, to turn their house into a seal hospital at a moment's notice, or care for a critically ill young female seal the day after returning from hospital with their newly born baby.

Of course, the conservation of all wild creatures hangs in the balance, as humans manipulate and manage and change more and more of the remaining areas of wilderness to suit their own needs. Perhaps we once thought the sea would escape man's grasping hands, but we have read too often of the devastation of animals and habitat caused by drift nets, oil spills and chemicals. There are no longer any safe areas.

Fiona's hands are tender hands, as she plays her violin; as she nurses and tends the baby seals; as she embraces her own children. And her heart is a caring and loving heart – one that erects no barriers. All will be cherished. I believe that the Fairy Queen who rules her kingdom on the Fairy Hill which looks over Cnoc bay, has touched Fiona with her wisdom and given her gift of music a different dimension. Although, in this book, we cannot hear Fiona sing, you will not fail to hear the cadence of her voice as she invites us to her island, to her home and to share with her her understanding of those vulnerable and beautiful 'people of the sea'.

Virginia McKenna
The Born Free Foundation
April 1995

Her stage was a granite rock, crept with lichen yellow

ONE

Islay Mist

Walking to centre stage, beloved violin tucked familiarly under arm, she scanned the audience. A sea of faces stared up expectantly from the depths. A murmur, a whisper; restlessness of unassuaged curiosity.

It was the hard-rock music of famous star Alex Harvey they had come to hear. The London Palladium in the early spring of 1978 was the historic venue for the Scottish musician's come-back concert. And here stood a 19-year-old girl, barefoot in a fairytale dress, lifting fiddle to chin and playing her soft, sweet melody – a Scottish 'Greensleeves' – in front of a mock medieval castle. But while she played, little did they know how her thoughts wandered.

Her stage was a granite rock, crept with lichen yellow, encompassed by an azure Hebridean sea. Thick, cool mist hung ethereal over the surrounding rocks and skerries. An inquisitive ensemble bobbed and porpoised, frolicking carelessly to the well-known resonance that spanned the ocean skies. The tune was born of the wind and the salt water, and the dedication was to the islands and those to whom she had first played her fiddle-tune 'Islay Mist' – an audience of wild seals.

A fantasy? A dream? No; these are my very own special memories. That magical fellowship was first briefly glimpsed, a chord struck by fate, when, as I played the violin on a lonely shore on the Hebridean island of Islay, two seals surfaced close in and lingered.

Knowing these creatures to be wild and free, then surely they would be wary and shy, and avoid human contact? I kept my pose but continued to play, observing them carefully without eyeing them directly. I sensed that the music held them there; the sonorous notes travelled the chill winter air to form a link that sustained this fortunate encounter. The seals' heads were shiny, dark and sleek, and their bewhiskered faces puppy-soft. But it was the eyes – those beautiful, wonderful eyes – that drew my attention. Ink-black, fathomless eyes embracing the depths of uncharted aquatic haunts. One seal quietly gazed at me, his body gently oscillating with the swell of the sea. The other glided past, polished head in profile, but his eyes, ever watchful, never left me. It was a steady, sidelong inspection. I caught a glimpse of the white of his eye before he gently slipped beneath the waves and rose again by the other seal a few moments later.

The tune ended and, lowering fiddle and bow, I looked straight at them. With one accord they dived like lightning, leaving a resounding splash and an upsurge of silver spray in their wake. Silence fell again; then two dots appeared far out in the bay. Somehow, I knew that although those eyes were still turned towards the shore, there was little chance the seals would return that day.

The grassy rocks on which I stood were the remains of a dùn, built strategically on a small promontory in ancient times. As the breeze strengthened and the waves grew choppy in the fading evening light, I reflected on how generations of seals must have swum in these waters and hauled out on the distant skerries over many thousands of years. I was yet to learn of the myths surrounding these beautiful animals – part of a relationship developed long through history between seal and man.

Immaculately designed for marine life, strong hind-flippers propel the seal's supple body while fore-flippers steer. With amazing speed, he catches his fish and can build up enough power to perform a whole series of porpoise leaps, which are enthralling to watch. He is comparatively slow on land – pulling himself along with fore-flippers and an undulating movement of his body – but an important part of his life is spent lying lazily on secluded rocks or beaches. A seal can dive to great depths and sleep upright far beneath the sea surface, 'bottling' every ten minutes or so to breathe, still dozy, before sliding back beneath the waves.

But looking back to that first fateful meeting, I realise just how ignorant I was. I knew so very little about seals, and their lives were a total mystery to me. Yet over many years of observing them, composing my songs and playing the violin by the shore, I have come to love and respect them. Living close to their raw, untamed world for so long, a bond developed between us in a very natural way. And when the seals came under threat from a terrible disease, the relationship expanded as, for the first time, I prepared to bring sick seals into my strange world to recover.

All this would never have happened without the encouragement and support of my husband, George, with whom I travelled to Scotland. It was with a sense of relief that I first stepped on to Islay soil. Our journey had been a historic one, made on the old *Viscount*'s last trip from Glasgow. The

Two seals surfaced close in and lingered (Chris Davies)

pilot had cheerfully told us this news over the plane's Tannoy and I couldn't help thinking someone had made a right decision. For 20 minutes before take-off I had been studying with great suspicion the strange collection of metal patches which were bolted, seemingly haphazardly, on to alarmingly bendy wings. However, this first visit to Scotland, in late November 1976, was to be an eye-opening delight and a wonderful turning point in my life.

Miraculously, the soon-to-be-a-museum-piece *Viscount* became airborne and carried us through the ethereal wintry skies across a beautiful landscape of highland, loch and sea. We descended beneath the final veil of cloud to discover Islay's scenic revelation lying quietly in the morning sunshine. Skimming across sparkling waves and measureless silver sands, the plane hopped over the sheep fence and bounded triumphantly to a halt on the desolate runway. Little did I know that 18 years later, Prince Charles, piloting a jet of the Royal Flight, would experience difficulties landing on this small Hebridean airfield.

The heir to the throne, having been a split second from death, walked unruffled from a terrifying journey's end. His plane lay like a drunken

I reflected on how generations of seals must have swum in these waters and hauled
out on the distant skerries

goose, nose down in peat-bog at the end of the tarmac, crippled by burst
tyres and millions of pounds' worth of damage. Island airfields were not
built to take sophisticated jets. They were designed for aircraft like the
trusty, if rusty, old *Viscount* and the boxy flying tractors which cover the

His plane lay like a drunken goose, nose down in peat-bog at the end of the tarmac

Islay run today. The prince toured the island and outside one of the schools, our daughter, Hannah, then aged six, excitedly handed him a card she had made especially for the occasion and told him where she lived.

The whisky distillery at Laphroaig had recently received the Prince of Wales Royal Warrant of Approval, and when he dined there, he was presented with two casks of whisky. One was a 15-year-old malt which was to be bottled with a special label and sold in aid of the Cancer Relief MacMillan Fund. The other, a ten-year-old, was to be stored along with the thousands of barrels in locked, bonded warehouses on the island. After five years it too would be bottled and sold for charity. However, after we had been settled in Kildalton for a few years, an incident occurred in which an enterprising islander made a copy of one of the warehouse keys, and a free flow of illegal whisky was rumoured to have found its way into many ordinary folk's homes.

It so happened that around the same time, my husband, George, and Roland our neighbour at the lodge, had built a stone wall at the estate entrance and erected tall pillars on which to hang a pair of impressive, heavy wrought-iron gates. It appealed to their humorous spirit to put a bottle of whisky in each pillar during construction; the idea being that unsuspecting visitors looking for a dram would be told that their host was just going out to the whisky well! A feed-pipe led out from each pillar with a tap at the end to facilitate extraction.

The famous Islay grapevine spread this story far and wide, so that when a squad from Customs and Excise arrived on the island to investigate what was becoming a serious leak, we were subjected to a dawn visit. Three stern-faced men in smart suits arrived in a Mercedes. Carrying officious-looking briefcases, they woke our household and asked to search the place. It took several hours to persuade them that neither we nor Roland had anything to do with disappearing Islay malt – even though Roland had an emptied whisky cask in his garden. He had acquired this whisky quite legally through a friend – Iain Angus, a bank manager on Islay, who collected barrels of malt as a hobby. When he sadly died, his widow was left with many hogsheads of Islay and Jura malt whisky in bond, with duty to pay as they matured. So, she sold the barrels on to friends of theirs.

To cheer up our visitors I sang my song 'Whisky Island', in English and in Gaelic, bringing a smile to the younger men's faces. But we got the feeling that this didn't go down too well with the most senior member of the group, who perhaps thought I was not taking their concerns seriously.

WHISKY ISLAND (UISGE NA BEATHA)

Islay shores, I've been longing to see you,
let the ferry speed me on my way,
Queen of the Isles, how I'm longing to see you,
Islay shores, I'll be with you today.

Whisky Island, I never can leave you for long,
and like a lover I'm coming home to you,
I'll walk through the heather on the green hills of Islay,
and look over the sea to the Mull of Kintyre.

I will walk through the woods of Kildalton,
watch the deer and the seals in the bay,
far away see the snow-covered mountains,
Islay shores, I'll be with you today.

Tha fadal orm gu t-fhaicinn Traigh Ìle,
'S am bàta Luath 'gam ghiulain a null,
Tha fadal orm a Bhan-Rìoghail nan Eilean,
Tha fadal orm gus an ruig mi thu.

Our seal patrol boat went out to check all was well (Chris Davies)

Tir Uisge-Na-Beatha, chan iarrainnsa t-fhàgail a choidh,
Is mar rìbhinn an gaol riut teilidh mì,
'S gabhaidh mi ceum troimh raointean gorm Ìle,
Is chì mi thall ud Maola Chinn-Tìr.

Prince Charles made his exit from Islay in a helicopter. As I pottered around the yard at home that afternoon, I heard a distant throbbing. As the noise grew louder, I realised that it was heading my way. Sure enough, with a deafening roar, the red chopper appeared low over the trees behind the outbuildings, flew quickly over the farmhouse and headed out to sea. With alarm I noted that its course would take it right over the common seal colony on the far islands, where mothers were nursing their newborn pups. Our seal patrol boat went out to check all was well, and no harm seemed to have been done on this occasion. But because of the potential disturbance at such a sensitive time, a letter expressing our discomfort was written. On our behalf, John Robins, Organising Secretary of Animal Concern, explained to the prince that he did not get our seal of approval for flying so low over the wild seal sanctuary. A sympathetic reply was duly received from His Royal Highness's Equerry. However, it is doubtful whether even the lure of a good Islay malt could bring the prince to chance piloting a plane back to this island after the worldwide publicity which surrounded his first attempt.

I was born in the village of Seal. Fate, it seems, had been weaving its web from my very conception, and in later life it was no surprise for me to learn from my husband that his mother's maiden name was Sealy. Seal is near Sevenoaks, a Kent commuter town half an hour's train journey from London. When old enough to earn a solitary freedom, I indulged my leaning toward the most remote places to be found in semi-suburbia. Woodland, cropped and grazed fields, overgrown quarries and an abandoned herb farm – these were my favourite places, my secret haunts, shared in bounding joy by Ninky, my beloved collie-Labrador cross. He would follow the scent of fox and badger, bother the rabbits and gently frighten the squirrels back to their trees.

At the age of eight, I learned to horse-ride at a school called 'Mrs May's', at Seal Chart, which was specially geared to children. I loved the excursions along quiet paths in this beautiful mature woodland. I longed to have a pony of my very own, but this was not possible. Therefore, I was overjoyed when I had learned enough to be allowed one of the school ponies to look after. I prepared him for rides; grooming his lovely palomino coat, picking out his hooves and tacking him up. I cleaned the tack after his day's work and rode Robin bareback to his field. It meant getting up very early on weekend mornings to catch a lift there, but I so enjoyed being with the ponies out in the fresh air of the Kent countryside that I did not mind. If someone had told me then that one day I would have my own horse, that I would ride along the seashore singing to my seals and canter through fields and mossy woodland where the deer roamed free, I would have thought it a fantasy.

The Sevenoaks area offered a wide range of musical activity. For me, this included a classical training on violin from the age of six and, from nine, I was taught piano with orchestral and chamber music in a wide range of ensembles. My sister and three brothers all learned to play the piano and my brother, William, went on to study piano and composition at the Royal Academy in London and in Krakow, Poland. My father had been a close friend of the parents of renowned cellist Jacqueline Du Pré and was her godfather. A competent pianist, who also composed chamber music which we could play as a family, my father's music was influenced by the Slavic gypsy style he had heard when touring the Balkans as a young man. William and I played the violin; Roy, the cello; Jean, the clarinet; and Nicholas, the oboe. Although my mother hardly ever played the piano herself, she was musical and was always ready to give good advice as we practised.

In our teens, William and I branched out into other kinds of music, especially the jazz of the 1930s and 1940s. I had great pleasure in working with a group playing jazz in the Hot Club de France style. While local jazz musicians the Townend brothers and a couple of friends strummed in the style of guitarist Django Rheinhardt, I attempted to emulate the wonderful violinist Stephane Grapelli. Playing every week along with many other talented musicians from the area was great fun. The 'Tepide Club' was held in a local pub and the evenings on which we met were always well-attended and lively with plenty of friendly banter. There was rock and roll and boogie-woogie piano, bluegrass and other folk styles.

I played at various venues with different groups, and I can recall with a wince what must have been the smallest concert audience in history. The bluegrass band had been booked by the council of a large Kent town to play one day in the park bandstand. But as we arrived the skies darkened, the heavens opened and a heavy thunderstorm proceeded to wreck the afternoon and our gig. Only a handful of very brave enthusiasts turned up with macs and umbrellas, while a few others hurried through the flooded park – obviously only coming out because they felt they should walk their dogs. This problem was not to occur again as my future audiences did not mind being wet at all!

I have enjoyed writing poetry from a very young age. In the year before my first visit to the Hebrides, I had singing lessons at West Kent College while studying A-level music and then I began tentatively to compose a few songs with the piano. The first time that I sang one of these was with a folk-rock band that I had formed with fellow students from college.

I returned to Kildalton several times over the next two years. I was living in a house near Sevenoaks town centre, working on demo recordings of my songs and liaising with an Artist and Repertoire man from Chappel's. It was during this time that I came to perform at the London Palladium as the opening act for Alex Harvey. Alex liked to relax in the bath at his London home in Fulham, listening to these home recordings of my early songs – a complete contrast to his hard-rock music. Sadly, he never visited Islay, though Ted and Hugh McKenna, drummer and keyboard player from his band, joined us on a holiday here. Alex was fascinated with our stories of Kildalton, the seals and the haunted castle – this undisturbed fairyland so far away from city life. It was he who asked me to compose a Scottish

'Greensleeves' and play it for him at his come-back concert at the London Palladium. Sitting on the beach, playing to the seals and walking in the woods at Kildalton, I composed 'Islay Mist'.

George, as my manager, had once asked me what I thought the pinnacle of my career would be. I told him that it would be to play the London Palladium. He then asked me if I would go with him to Islay and live with him there permanently once I had achieved this. I agreed, never dreaming what the future held in store. It was the year that Scotland's football team had earned a place in the World Cup in Argentina. Dave Boyle, a Scottish songwriter and guitarist who had spent time in Islay with us and had worked with me in my studio in Sevenoaks, wrote a rock football song for the team called 'Royal Blue'. I wrote Scottish fiddle parts for the song and composed a jig for the B-side – 'Hoots Boots'. Chappel's took the song and my jig and, with producer Derek Wadsworth, we prepared to record at their professional studios in Bond Street.

The Sensational Alex Harvey Band played the music along with a few session musicians and Alex was keen to sing the lead. Unfortunately, there were problems over Alex's recent split from the band and it had become a serious legal dispute with his former management. When Alex's new manager heard of the plans for him to sing a football song, George was summoned to a dingy basement office off Edgware Road. Dave Boyle, cousin of the reformed but infamous Glasgow hard man, Jimmy Boyle, had warned George to be careful if he encountered Alex's security man, John McKillop. Ex-army, McKillop was over six foot, wore leather and had a reputation for being unpredictable. George sat at a desk opposite Alex's manager and Alex hovered at the side amongst recording equipment.

To George's consternation John McKillop appeared and stood right behind him. On the basis that caution is the better part of valour, George agreed with everything Alex's manager was saying. He acknowledged that perhaps it was not such a good idea for Alex to sing the lead on a football song with the band, considering all the legal problems. Yes, of course, he understood that it was not even the right image for Alex. George even apologised for not dealing with Alex through his manager.

However, when he was told I was not going to play 'Islay Mist' at the London Palladium, George stood firm. There was a long silence and he felt his neck prickle. But, at last, I was offered the chance to play – provided I proved myself at rehearsals. Afterwards, George explained the difficult situation in which he had been placed, but he preferred the attention of the security guard to the wrath of a heart-broken nineteen-year-old girl, with his hopes of a future on Islay with her crushed.

It was only in later years that I realised that George had taken quite a risk. We saw on the news that McKillop, now calling himself Miller, had kidnapped Ronnie Biggs, bundling him into a bag and speeding him off in a powerboat. George says that if he had known this at the time, I would never have played at the London Palladium because he would never even have gone to see Alex's manager. Fools go where angels fear to tread.

John McKillop helped to build a castle for the stage. To Alex and ourselves it was Kildalton Castle – even though it had more of a medieval appearance to fit in with the 'Camelot scene', as the *New Musical Express* described it afterwards. The rehearsals went well with McKillop practising a sword-battle routine with another stunt man. They were to fight over me; a good white knight and a black evil knight – who, to Alex, represented his old management company. As the black night was vanquished, Alex, as the king, was to make his own spectacular entrance by leaping over the castle parapet.

On the big night, the mood of the audience was buoyant as I played 'Islay Mist' at my dream venue and we performed our play to introduce Alex. However, the original intentions of the stage director went hilariously wrong. As George was eager to be involved, he had been given the apparently simple task of turning on a tape on the huge control console of various recordings of horses' hooves. A steady clippety-clop coming to a halt was required, but George wound the cassette on to the right number, not realising that someone had zeroed it the night before. On the cue from one of the stuntmen – 'Hark! I hear a horse approaching' – George pressed the play button and the Charge of the Light Brigade thundered through the London Palladium! There was much laughter from the audience – they probably thought that it was intended. Then Alex sprang on to the stage from the castle walls. The curtain rose on his new band and the orchestra, conducted by Derek Wadsworth, struck up as Alex swung into his first heavy-rock number.

During my first short holiday on Islay, a special love for Kildalton had been kindled that would grow and mature as it became my permanent home and my inspiration. On that first car journey to the small hotel on the estate, some ten miles from Glenegadale airstrip, the wide range of Islay's terrain became apparent. The uninhabited peat-bog stretched far on each side of the coastal airfield, the hills appearing low in the distance. On the road 'up country' from Port Ellen, gently rolling farmland met the rocky seashore.

Soon the island of Texa breaks the sea view and then the road winds through the whisky distilleries of Laphroaig, Lagavulin and Ardbeg. These imposing buildings, with their fascinating pagoda-shaped roofs and accompanying clusters of white-painted cottages, export Islay's water of life to the farthest corners of the globe. By the time I reached Kildalton Estate, with its ancient forest, small secluded bays and a myriad of scattered islets, much of that first desolate impression of Islay had been dispelled.

In those first weeks I began to play Scottish fiddle music, jigs, reels and airs. The Dower House stands alone by the shore of the bay beneath the Fairy Hill. With the help of the dining-room piano, my first Islay-inspired songs emerged. I climbed the broken staircase of Kildalton Castle's tower and played across the crisp air of the deserted wood, over the empty sea, to the snow-covered hills of Kintyre. I stood on the dùn – a rampart from a dark and distant age – fingering melancholy tunes in the still air, bringing the seals to me for the first enchanted time. Climbing the Fairy Hill, I gazed upon herds of deer grazing peacefully in the glens, saw shining peat-lochs reflecting the mood of Hebridean skies and shared the eagle's view as he surveyed his world in all its rugged glory.

George, whose unerring love and vision for Kildalton's wilderness and wildlife have sustained and educated me, bought the farmhouse which has been my home for the past 16 years. The extensive outbuildings and the grounds became accommodation for ponies, milking cows, ducks and geese, hens and peafowl. Since then, however, the stables have housed sick deer

From the farmhouse, the land sweeps down the wooded slope to one of the most sheltered and beautiful bays on the estate

Deep abyssal eyes incited the myth that they bore the souls of the ancient dead

and injured seabirds. Even the en suite bathrooms of the farmhouse, which had a brief career as a hotel, have become temporary homes to sick and injured seals. I have raised a family of swallows in my bedroom and coaxed torpid bats and car-stunned owls back to mobility.

From the farmhouse, the land sweeps down a wooded slope to one of the most sheltered and beautiful bays of the estate. Here the sea is safe and shallow with a sandy beach, but, on each flank of the bay, the shore is stony and rises to small cliffs topped with impenetrable rhododendron. To one side, a path winds down through the trees from the castle to arrive at a small pier, and a collection of rocks forms an island in the centre of the bay. This jewel in the Queen of the Hebrides' crown was to become of great significance when raising abandoned seal pups.

I learned the ways of the seals, discovered their favoured music and composed songs while honoured with their friendly presence. In peace and serenity we communed, our secret affinity mysterious and unfathomable.

23

Deep abyssal eyes incited the myth that they bore the souls of the ancient dead. I learned to sing in the Gaelic tongue – the ambience interbred with this tousled but placid wilderness.

For two years, I never left the island. Although I had much to be grateful for in my upbringing, I had no desire to return to 'civilisation'. I loved my new home and way of life at Kildalton.

CITY LIGHTS

City lights seem so far away now,
trains and fast lanes and rushing around,
here am I in my own paradise,
living dreams and loving you.
So far, seems so far away now.

Country girl never liked the city,
always dreaming of valleys green,
island shores I'm so far away now,
with deer in the hills and my love for you.
So far. So far.

City lights fade away now,
you don't shine, shine for me now,
I walk alone and watch the sea,
my love for you turned the tide on me,
city lights fade away now.

All alone by the rocky seashore,
my green valley in woodland wild,
here am I in my own paradise,
living dreams and loving you.
So far. So far.

City lights seem so far away now,
city lights you don't shine for me now,
you can light up the sky above you,
you're so bright, but I'll never love you.

TWO

The Fairy Hill

There was once a Danish princess who formed the islands between Ireland and Loch a' Chnuic, the Bay of the Hill. Carrying an apron full of rocks of varying size, she scattered them as she went, creating the islands of Rathlin and Texa, and the string of smaller isles and skerries off the shores of Kildalton. When she reached the bay below the Green Hill she died exhausted in the shifting sands. She was called Ile, and this southernmost island of the Inner Hebrides is in legend named after her. Buried in a hollow on the side of the hill by a hairpin bend on Seònais brae, she commands the view across Loch a' Chnuic to these isles. Today, two standing stones mark her grave; lying far apart, as she was a giantess.

However, it is the little people who have given this conspicuous hill its name. It is the Fairy Hill from where the Queen of the Fairies rules. Rowing in from the farthest corners of her kingdom in their tiny coracles, the fairies pass the skerries where watchful seals lie like sentinels, guarding the way to a fairy palace beneath the rocks and green turf of the hill. They enter Cnoc Bay and, beaching their little boats there, they make their way to a concealed entrance on the grassy slope. Humans, too, were once welcomed at the throne of the Fairy Queen. From a magic cup she dispensed wisdom to all the clans of the land. Some were late in coming, however, and so remained dull-witted. It is thought by some that to enter the hill, a human must run around it backwards six times during a full moon at certain times of the year.

There are many stories of beautiful music heard coming from beneath fairy hills. It is also said that they like to lure human musicians to their dwelling places. One tale has it that two travelling fiddle players were approached in a lonely place by a little, white-bearded old man in strange attire. He asked them to play for his friends and offered a handsome fee. Having agreed, they followed him to a rounded hill, the entrance of which opened at his command. Inside a brightly lit cavern, the little people dressed

Buried in a hollow on the side of the hill by a
hairpin bend, she commands the view across
Loch a' Chnuic

in green were banqueting and making merry. After eating well, the fiddlers played jigs and reels as the fairies danced throughout the night. At dawn they were paid well in gold but, on returning to the town, they were shocked to find that everything had changed and 100 years had passed – a common theme in stories of humans entering these hills.

Although it appears that fairies live in a different dimension to us, perhaps music can transcend the intangible barrier. Ancient lore holds the seals themselves to be fairies. Who, then, but the Fairy Queen herself should draw me into her domain to serenade her subjects?

There is another grave upon the Fairy Hill, that of John Talbot de vere Clifton. He owned Kildalton Estate in the 1920s. In those days the estate included many thousands of acres to the north and south of the present land, and he was Lord of the Manor at Lytham, Lancashire. Clifton was a restless

OPPOSITE:
It is the Fairy Hill from
where the Queen of
the Fairies rules

26

There is another grave upon the Fairy Hill

traveller and, before he was 20 years old, he had been twice round the world. He later went to North America and Canada, exploring the remote wilderness by dog-sleigh and kayak. Journeying with the Indians and Eskimos, he hunted the musk ox and the caribou and named lakes and mountains where no white man had been before.

However, he loved his Kildalton home and delighted in the wildlife. After many years of playing the violin to the seals, I was amazed to discover that I had a predecessor in the intrepid laird. He had a silver flute which he took with him on his travels, and a golden one which was kept for playing at home. When he played at Kildalton, the seals came up near the rocks where he stood and listened to the music. Later, I was perhaps not so surprised to learn that this fearless adventurer would play games with his daughter in the woods of the estate, sharing her love of the highland fairies – and even leaving 'elf-written' letters under the trees for her to find.

On his death in Tenerife in 1928, Talbot Clifton's body was shipped home to his beloved Islay for burial. He had spurned the family vault at Lytham where generations of squires were buried, telling Violet one night that if Africa killed him, he wished to be brought home to Kildalton to rest on the Green Hill, where the deer fed and the birds flew in from the north. His embalmed body lay in Kildalton Castle until friends and family had gathered for the Highland funeral.

The cortège followed the horse and cart that took Lord Clifton to the foot of the hill, then strong men carried him to his chosen place halfway up the green slope, looking out to sea. They lowered the coffin, made of lead and with a glass top, into the grave. The laird's wife threw in a wild violet and a bunch of orchids, and down from the high hills an eagle soared, its shadow falling on the coffin as he circled twice and swooped low.

Clifton's grave is marked by a cairn and a cross. Bowls of stone collect rainwater for the deer to drink, and in spring the daffodils nod their yellow bonnetted heads in the sea-borne breeze.

So still was the air on that winter's day when I first climbed to his resting place with George. Such quiet, such all-pervading peace; the silence stating its presence like a living being. Passing the sparse ruins of some old

The Celtic form of my cross echoing that of the cross high on Clifton's grave

cottages, we ascended from the scrub of ancient oak woodland at the foot of the hill as the view unfolded exultant before us.

A bleak land, fragmented with patches of snow, stretched back towards the far heights, while below us shone a pure, glassy sea. Palest of blue and grey, fulfilled with brilliant light, the oceans met the firmament and communed as one. Beyond the idle bay, among far skerries and small islands, the seals were yet concealed from me. Embraced by thick rhododendron, the shoreline of little bays led from the Dower House to surround a peninsula clothed in a mantle of woodland. To the southside, the land sloped gently to the sea. Here it continued as a treeless, tidal island encompassed by a multitude of rocks and islets – another playground for my future companions. In the large bay by the road, known as Seal Bay, I first set eyes on a wild seal as he lay on one of these little isles.

Its wizened branches, cruelly twisted by gales, stood bravely against the magnificent scenery

Further still from distant Kintyre's shores, rose resplendent snowy peaks – the mountains of Argyll and Arran, and to the north, the three Paps of Jura. For a time the rocks and low islands beyond the bay of the hill, bathed in a golden light, seemed to gleam with an inner radiance. Kildalton forest lay motionless on the peninsula, its hidden depths unruffled, secret. But, from this sombre reach, dark as the winter timber, a castle tower topped the trees to lie in the path of the entombed laird's gaze. And here, alone in a room bedecked in black, she kept a vigil for three months. A widow, mourning her husband. Violet grieving for her beloved Talbot. It was her fervent prayer that the hill remain unspoilt, so that, at the miracle of her husband's resurrection, his view across the islands and the sea would be unchanged.

From the Celtic House (Roy's shop in Bowmore), George had bought me a large pewter Kildalton Cross on a chain. I wore it round my neck as we wandered about holding hands on the Fairy Hill; the Celtic form of my cross echoing that of the cross high on Clifton's grave. We climbed the steep incline to the top of the hill, examining the trig point, the old stump of a flagpole and the heavy metal rings set into the ground on which cannon were once mounted.

On the larger island of Churn stands a lighthouse like the last outpost of man in a foreign land

As we walked, George talked with fervour of his love for the estate and his hopes for the future, now and then pausing to take a photograph with his precious Nikon camera. The beauty of Kildalton and the freedom of its wildlife had to be preserved, especially at this time when it was under threat from a massive development planned by its owners. Together we sat in the stillness, in hushed contemplation. There was a small old tree standing alone – the only one to have grown and survived at this level. Its wizened branches, cruelly twisted by gales, stood bravely against the magnificent scenery. Deserving recognition, we named him Archie, and our tree was to be fondly visited many times before he finally fell, succumbing to the fierce winds. As we looked across the sea to the string of isles far beyond the Dower House Bay, I saw them as empty, even desolate, little knowing how the creatures inhabiting them were to enrich and shape our future.

Taking a picnic in the boat with her, Violet Clifton would visit these little islands. Land entranced by the sea – the seals' domain. Ruled by the state of tide and wind, they lie mellow and welcoming in the light summer breeze, becoming harsh and inhospitable when dashed by fierce gales and unbridled seas. In summer, the rocks are coloured brightly with clusters of sea pinks and yellow lichen, gently toned with greys and greens. Larger islands nurture a thick sward of wild grasses and flowers atop a jumble of huge boulders. In winter, white horses chase wildly from the open ocean to crash relentlessly on the dispassionate stone.

In calmer seas I have often taken the journey from the warm shelter of the peninsula out to this in-between otherworld across the bay. Leaving small cliffs restless and noisy with breeding seabirds, the boat heads out to the serenity of my favourite auditorium. Beneath the waves are many hidden reefs which are so dangerous to seagoing vessels that the Admiralty Charts class the area as 'foul'. On the larger island of Churn stands a lighthouse like the last outpost of man in a foreign land. Apart from a few small yachts and creel fishing-boats, the seal colony is left in peace.

The audience awaits, some dozing yet watchful, others idly fishing. For several months during the spring and summer, these isles lie awash with seals. I pass by many smaller rocks laden with their sleepy bodies. Some lift their heads to scrutinise the intruder while others scratch on neck or belly with a flipper; nonchalant, sun-soaked.

All the seals are dark in the water but, on drying, the true colours of their coats are revealed. The common seals that colonise Kildalton shores have a wider colour range than other colonies. From dark grey, flecked and spotted with white, through sandy colours to pale grey and white, each seal is quite individual. Some are juveniles, their behaviour particularly playful, while others are not so young, their whiskers stout, with large, wise old faces – for these seals can live for 30 years.

Slowing the boat, I enter 'the river', a channel dividing the beautiful Eilean Bhride from Ceann nan Sgeirean. Here, with my picnic, I can sit quietly on the grassy turf a little way inland, and watch the comings and goings of seals through the gap between the islands. I have watched mothers with pups riding on their backs, playing together in the shallows, and nursing and slumbering contentedly on the shore – always inseparable. On a higher tide, as the seals take to their fishing, I'll take to my fiddle and make for a comfortable rock by the sea. And, as I play, the seals dance, bobbing and leaping, diving and rising again. Some will come out from behind other rocks, straining higher in the water for a better view. Others stay on the receding skerries, loathe to move until the tide covers their bodies, stretching their supple necks high, hind-flippers curled towards their heads. By some sixth sense a cavorting group swimming near me will submerge as one with a splash, then slowly emerge, seal by seal, to gaze again.

I can only imagine their antics under the sea in deep water, but there are many other rocks to play from. In sheltered pools, reflecting the deep blue of the summer skies, the water is clear enough to see to the sandy seabed. Looking down from the drifting boat, or from rocks pebbled with limpets and winkles and strewn with wrack, an exquisite aquarium can be seen. Here and there beds of seaweed, eelgrass and mermaid's tresses cover the sand. Like a marine garden, the rocks are decorated with strawberry anemones and pale rose-pink urchins, while little movements here and there catch the eye – a golden starfish, crabs of all sizes, green and brown; sometimes a skulking lobster. And here the beauty of a seal's movement underwater can be savoured. Gliding smoothly, master of his liquid home, twisting and turning, fluent in his own element: choreographer of an underwater ballet.

To the north of Eilean Bhride lies Outram, an island surrounded by scattered rocky isles, some of which encircle a large pool. There is an entrance from the open sea which also leads to inland waters. Here the seals can be sure of very little disturbance and, since those that are slightly injured come here to haul out in peace, we have come to know it as the Seal Hospital. From here they have not far to swim to the fishing-grounds further out to sea, where sometimes a shoal is marked by a pod of turbulent seals, the happy gathering leaping and diving as they feast.

As I return to the peninsula, like a beacon the cairn and cross of Clifton's grave gleams white on the Fairy Hill. From a boat in the bay below, on one warm, calm summer's day, I scattered the ashes of my father. For he, too, loved the quiet unspoilt places, the beauty of God's creation.

And it was on this sandy shore of Cnoc Bay that a sick seal pup was found, abandoned and malnourished, washed up under the trees. He was the first that I cared for, and we called him Tarzan.

The last light of the day was melting into a creepy gloom as George and I left the Green Hill that first winter's evening. Returning to the quiet hotel, we were surprised to find out just how long we had been. It had seemed but a short while, but the best part of a day had passed, as wandering in that fairy place the bonds of time yielded to the tenacity of our new love.

On our return from the hill, George's camera became faulty. The shutter had jammed and it was with great difficulty that he removed the film. On my first visit to Kildalton I knew nothing of its past. Through history those with authority have chosen to settle here. The area, where in legend Princess Ile is buried, is an ancient site known as Clachan Ceann Ile, which can be translated from the Gaelic as 'Township of the Chief of Islay'. In ancient times it was believed that fairies were the spirits of the dead and that they could influence the affairs of man. As my knowledge grew and events unfolded, I came to accept that this is indeed a mystic, powerful place, where strange things can happen. The electrics on cars cut out inexplicably – often, to the driver's consternation, at night in the wood around the castle. Local people speak of the Kildalton Curse and many fear to walk here after dark.

A young boy was believed to have met Clifton on the hill, saying the person he saw was 'not a right man'. It is said that this laird is seen before a death on the estate. And there is a tale of a resident who, while driving at night on the road near the hill, ran over a man who suddenly appeared ahead of him. He was so convinced he had killed someone that he searched, vainly, for a body.

Not long after I first came to Kildalton there were two Irishmen working at Ardbeg distillery on a short-term building project. Hearing from locals the story of Clifton buried in a glass-topped lead coffin, they decided to dig down to his grave for a look. As they were busy removing earth, a man in strange clothes appeared from behind them and asked what they were doing. He took them to a thatched cottage on the side of the hill and gave them a dram. The atmosphere turned chill, and they felt they were in another world. White and shaken, they arrived at the Dower House Hotel and told their tale, whereupon they were assured that the house they had described did not exist. Absolutely terrified, they left the island the next day and have not been heard of since. The day following their adventure, George went up the hill to see what they had done. There was a narrow but deep hole in the grave. George felt he should do something in apology for this desecration so he put a glass with a dram in it down the hole, then filled it and tidied up the ground. Before leaving he placed daffodils on the grave.

Animals often seem to sense things that we cannot. One day, my stepdaughter Emma and I thought that it would be a good idea to ride our ponies, Misty and Jason, up the hill. We took the route to the grave, planning to continue to the very top. As we approached the cairn, both ponies pricked their ears forward and seemed to be aware of something we could not see. They halted and refused to go on. Although I saw nothing, I had a strange feeling, as though the atmosphere was charged. I have visited the grave quite happily many times before and since, but on this occasion it seemed prudent to leave the hill. I never rode up there again. Clifton loved horses and was internationally renowned for his skill in training them and the special rapport he had with them.

In the days that followed our walk on the hill, I took my violin into the derelict castle and climbed the dangerously broken staircase until I reached the top of the tower. I had passed the room of Violet's vigil, where scraps of black wallpaper still clung to the plaster. From the open ramparts of the turret, a magnificent view took in the surrounding woodland, the fields of the home farm, and the sea beyond the trees. Across to the Green Hill I played gentle melodies, Talbot's grave clearly visible, a cairn shining in the cold winter sun. As I returned my fiddle to its case, like an echo the curlew called; a piercing cry through a tranquil landscape.

I followed a lush path down to the shore. On either side the old trees stood tall, their leafless limbs damp and green with moss and lichen. I stopped to play again where the rocks rise from the sea and are met by those which were shaped long ago by man. And the seals who came – were they descendants of those who delighted in the golden flute? And later, as I composed a song of love, did I waken the passions of those who sleep?

I knew nothing of the history of Kildalton, but after wandering on the Fairy Hill and playing across to the seal islands and to the hill from the castle tower, I wrote this song at a piano in the Dower House.

THE FAIRY HILL

Our love wanders on the Fairy Hill,
echoing the music of our lost hearts.
Across the isles and the sea,
to the mountains and the sky,
and back again.

Our love, answering a distant call
with harmony,
and there, hand in hand,
strolling, running, smiling, loving,
we remain.
Far away from time.

Our love, longing for a destiny,
searches for a chord of light
to carry us across the sea,
where the curlew cries, through the mist
to mountain shores,
to learn again.
To be free.

Our love wanders on the Fairy Hill,
echoing the music of our lost hearts.
Across the isles and the sea,
to the mountains and the sky,
and back again.

On returning to Sevenoaks, George helped me set up a small studio at home. Accompanied by piano, violin and synthesiser recorded track upon track, I produced 'The Fairy Hill'. The music was ethereal, but I was not prepared for the strange effect it seemed to have. As I played it while I relaxed with George one evening, the room went suddenly cold and a glass shattered spontaneously. I took the tape to Kildalton. The farmhouse in which I now live had been partly converted and was at that time Kildalton Hotel. One evening, after the bar had been shut up and locked, we were sitting in an adjoining sitting-room with the barman and a few others. To entertain the company I put on a tape with some of my songs. As 'The Fairy Hill' was playing we were all startled as the sound of crashing glass came loudly from the bar. Convinced that a shelf had collapsed or some such mishap, the barman opened up the room to investigate. There was nothing out of the ordinary, nothing broken. The bar-room had been converted from Violet's private chapel.

I played the song on Kildalton once more before deciding it was best laid to rest. On this occasion George and I were with our neighbour, Roland, in his extension at Quartz Lodge. A lawyer and his wife had joined us and we were having an enjoyable evening. Again I put on the tape with my songs. As 'The Fairy Hill' played, there came loud scratching noises from the bushes outside the closed window. Perhaps it was the wind? But no. Whatever it was came through the window and on to Roland's desk where the papers rustled and moved. The atmosphere in the room was electric. As the song ended, peace was regained, but we were all shocked and the lawyer's short hair literally stood on end. It took much comforting and persuasion before he would leave the house, having to walk in the dark to his car.

Quartz Lodge, situated at the entrance gates to the heart of the estate, was built in beautiful, white-marbled stone as a wager. Around 1900, a disagreement arose between a professor of geology, who had a survey report which stated that there was little quartz on the estate, and the laird, who insisted this was wrong. He duly had Quartz Lodge built to prove it. In the

The aspect of the gable end became more and more askew

early 1970s, a travel article in the *Sunday Telegraph* described how the writer had come to Kildalton and found, nestling in the woods, a cottage built of quartz, and, the article went on to say, you would have sworn that Snow White paid the rates.

However, when Roland had an extension built, it gradually began to subside; cracks appeared inside and the aspect of the gable end became more and more askew. This was highly amusing for everyone except Roland. He had to sue the contractors and eventually a new extension was built. This time, however, it was decided that, as a precaution, the Daoine Sìth should be taken into consideration. It was a possibility that a path used by the little people on the way to the nearby Dìgh, the Fairy Hill, had been built over. To prevent them venting their wrath on the construction, a fairy subway, made with a large pipe and little doors at either end, was incorporated into the building plans. The extension stands straight and true to this day.

Those I play to and heal are descended from the seals who first colonised the rocks and skerries of this coast in the distant past. Truly, this is the kingdom of the seals

When the more remote parts of the island received electricity, hydro-poles were erected. The route chosen for the cable going up country made it necessary to put poles up on the Fairy Hill. Unfortunately, one of these was placed in a position which spoilt the view from Clifton's grave. Inexplicably, it fell down, was re-erected, but fell again. The Hydro-board was forced to re-site the pole. As part of the planning requirement for the proposed large development, a water survey was undertaken. An apparently fit and healthy man went up the hill to take measurements of the water flow. He stood in the burn and dropped down dead, which hindered progress of the plans. This burn is called Allt a' Chrochaire, which means Hangman's River.

From the remote days of their birth, through the elusive mists of many ages, the shadows of history endure in tales of myth, lore and legend. Enhancing what is known and understood with a blend of the possible, the improbable and the unbelievable, these stories breathe life into the inanimate records of the past. Remnants of our lives may become legends of future generations. Islay is blessed with a wealth of inherent tradition. Kildalton's unique sentience is only fully perceived when an appreciation of its beauty is enriched by knowledge of its history and folklore.

And so it is with the seals, for they have lived here alongside man unchanged through many ages; a common thread running through Kildalton history. Those I play to and heal are descended from the seals who first colonised the rocks and skerries of this coast in the distant past. Truly, this is the kingdom of the seals.

ISLAND SPIRITS

Down by the water,
the sun in my soul,
toes in the sand and a hazy horizon.
A whisper of yesterday's heroes,
my island.

Deep in the forest,
beyond in the ferns,
feeling the future the spirits are waking.
Along forgotten paths remember
the island.

And if you think you'll make me happy,
show me how to touch you,
and live alive the dreams that you once lost.
But when the summer sun is over,
will you leave your lover
to weep among the shadows of the past?

I feel you near me,
saying you need me,
whispering in my mind.
Wandering only,
timeless and lonely,
what will your lost hearts find?

So many stories,
the legends of old,
tales of the norsemen,
the laird in his castle.
Along forgotten paths remembers
the island.

And if you think you'll make me happy,
show me how to touch you,
and live alive the dreams that you once lost.
But when the summer sun is over,
will you leave your lover
to weep among the shadows of the past?

I feel you near me,
saying you need me
whispering in my mind.
Wandering only,
timeless and lonely,
what will your lost hearts find?

I'll live alive the dreams that you once lost.
I'll live alive the dreams that you once lost.

THREE

An Lanndaidh

George was taking me on a mystery tour. There was something special he wanted to show me. We were in the thick mature woodland behind the castle, where the main path runs down a wide grassy glade to the sea. Before we reached the bay, he led the way up a slope to one side, where we were met by thick rhododendron. Ducking down under what seemed to be an impenetrable bush, George disappeared. Following him on hands and knees I discovered we were on an old muddy path, dank with rotted leaves and, for the most part, only negotiable on all fours. Standing up, we picked our way down a slippery incline and veered to the left. Here there were large trees which had fallen over the path, with more dense rhododendron on either side. Everything seemed dark and damp, with only a glimpse of grey winter sky visible through the sparse tops of old pines. Awkwardly, I climbed over the slimy trunks that lay in our way; spiked, broken branches catching on my clothes.

Then the cross came suddenly into view. Broken at the plinth, it lay at a crazy angle, held by the half-fallen trunk of a huge tree. In mutual support each had prevented the other from crashing to the ground. Around them the relentless invasion of rhododendron had continued, threatening to envelop, conceal. Still impressive in its near demise, this was a full-size replica of the Kildalton High Cross.

The original cross can be found in the grounds of the old Kildalton Chapel on Ardmore Peninsula, a couple of miles north of the forest where the replica was placed. The chapel dates from the thirteenth century but services have not been held there since around 1690. Its remains, roofless, yet with walls, window arches and gable ends still intact, stand alone on a sparsely wooded hill. Carved stone slabs still lie in the rough turf of the surrounding graveyard which is encompassed by a wall.

Hewn in one piece from local bluestone, the Celtic Ring Cross is magnificently adorned with ornate carvings. On one side are intricate

The original cross can be found in the grounds of the old
Kildalton Chapel

designs of interlaced serpents, said to represent eternity. On the other, biblical scenes are depicted, such as Abraham about to sacrifice Isaac and the Virgin and Child with attending angels. Nine feet high and uniquely well preserved, the cross is related to the Iona group of major crosses – St Oran's, St John's and St Martin's.

There were Viking raids on Iona in the late eighth and early ninth centuries. Many monks were killed, but others managed to escape. Some went to Ireland with 'the great gospel of Colum Cille', four highly illuminated gospels – their most precious relic – which were to become the

Book of Kells. Other monks reached Islay and settled in this south-eastern area where they made the beautiful Celtic Ring Cross. Kildalton can be translated as 'fosterling' – the fosterling church of Iona.

In 1882 the Kildalton Cross was tilting dangerously and Mrs Lucy Ramsay, wife of the laird, had the monument excavated and reset. The opportunity was taken to have casts made in concrete – one is now in the National Museum of Antiquities and another stands in the wooded grounds of the castle. However, when the cross was removed, they made a gruesome discovery. Underneath were buried the remains of a man and a woman. Their joints had been dislocated, indicating that it was most likely they died at the hands of the pagan Vikings, who tortured their victims by spread-eagling. This was a grim reminder of the events that had led the Ionan monks to flee to Islay in the first place.

Looking north from the chapel the three Paps of Jura rise majestically above the surrounding hills. Each year, at the end of May, the Bens of Jura

The Seal Islands, where the common seal colony breeds, lie off the peninsula on which Kildalton Chapel is built

race is held, when over 100 runners arrive from all over Britain. The races were held without any local runners until George and Roland formed the Kildalton Amateur Athletic Club. The club's members all wear green running-shirts with a bright orange Kildalton Cross emblazoned on the front. Always ready for a challenge, they could not see why it would not be possible to complete the course with planned training. For reasons of safety it was a rule of entry that runners first had to prove their running and orienteering skills on mountains such as Nevis and Snowdon, so difficult was the course. However, George approached the organisers and after discussion the rule was waived for locals who had trained on the Paps and knew the area well.

There are stunning views across Islay from these heights, and fierce west winds unleash their wild fury unhindered from the Atlantic. In the spring, George would travel to Jura to train on the Paps. From these impressive summits he enjoyed incredible views of Jura and of Islay across the Sound; alone but for the eagle and the shy mountain hare, who would crouch on the snowless rock, not understanding that his yet white-patched coat was no camouflage for him now.

Deer and rabbits would look up from their grazing as George ran past the castle and through the forest to Quartz Lodge. Turning into the single-track public road, his training route would take him past the Fairy Hill and down the slope to the bay. As George ran past the sandy beach, seals would raise their heads from rocks or pop up from the waves to see this human acting so strangely. Turning up the steep brae where Princess Ile is buried, the silver gleam of their reclining bodies would catch his eye. On the long road to the cross he would often see a red stag or a group of hinds watching him curiously from rough ground nearby. Sometimes a pheasant would be caught unawares, the startled bird complaining noisily as it made for the cover of ancient oak woodland.

These runs were exhilarating for George who discovered An Lanndaidh in its many different moods. Running in all weathers, whether straining against the bitter winter winds or cooled by a soft spring breeze, he came to know each rise and fall and bend in the road, every view of sea, loch and hill. He would repeatedly run up and down the steep north-west slope of the Fairy Hill and, as his training progressed, he would run from our house to the top of Ben Bheigeir and back – sometimes at night as snow lay on the ground and with only the light of the stars overhead.

George believes that it is one's inner being which masters the physical pain barriers that he has to pass beyond in order to progress in his training. The combination of heightened awareness and understanding of his own body and mental capabilities, coupled with the beautiful area in which he trains, creates a highly spiritual experience.

To George, An Lanndaidh is like an open-air cathedral. Unspoilt, the magnificent landscape and abundant wildlife are the power and the glory. Here is one place in the world where His creation can be seen in all its diverse splendour, and spending time quietly alone in this paradise on earth can bring one to enlightenment.

AN LANNDAIDH

There's a place that I know in the western seas,
where the world sleeps still
under the ancient trees.
I've heard it called by the name An Lanndaidh,
and there's a place, and there's a feeling there for me.

And there's a place that I know where I love to be,
sitting in the arms of an old oak tree,
looking overland to the old castle walls.
There's a place, and there's a feeling there for me.

Up in high hills and down by rocky shore,
shady woodland and sandy bay,
the swallow, the eagle,
the swan and the seagull,
awake in the dawn to live another day;
till darkness falls over An Lanndaidh.

If I walk through the woods to where I love to be,
sitting on a rock looking out over the sea,
wind, rain or shine, all beauty there is mine,
there's a place, and there's a feeling there for me.

There's a place that I know in the western seas,
where the world sleeps still
under the ancient trees.
I've heard it called by the name An Lanndaidh,
and there's a place, and there's a feeling there for me.

45

The Seal Islands, where the common seal colony breed, lie off the peninsula on which Kildalton Chapel is built. Although it is mainly the common seal that lives off this coast, individual greys, prolific on the west coast of Islay, are often seen mingling with this common colony. Both species of seal had long been part of the human economy in the west coast of Scotland, although the greys bore the brunt of the hunting as they become vulnerable when they come on land to breed. It is possible that the Columban monks would have used seal oil for their lamps and also availed themselves of the meat and skins. The seal oil-lamp, which has a rush wick and is known as a crusie, was used until the end of the nineteenth century. Sealskins were, among other things, made into boots until the introduction of rubber.

There is a wealth of folklore and superstition associated with seals in relation to hunting. A strong tradition prevailed that anyone who kills a seal will have bad luck until the day he dies. It would be a mistake to link stories of mermaids and mermen with those of seals. Numerous reported sightings of these creatures describe them as having a human torso with a fish-like tail, quite different from a seal's hind-flippers. The seal legends centre on the belief that a seal can, in certain circumstances, change completely into a man or woman and walk upon the land. They are said to have relationships with humans, often producing offspring with webbed hands and feet. Sometimes the baby is left with the human mother and she weeps for her husband who has returned to the sea. In other stories the seal mother, who always yearns to return to the ocean, eventually leaves her husband and child. The youngster grows up like a normal human, his webs having been cut off at birth, but he and the generations that follow him are inherently knowledgeable about the sea.

In the famous Orcadian ballad 'The Selchie [grey seal] of Sule Skerrie', a maiden falls in love with a seal-man and marries him. They have a son, but after a time the father returns to Sule Skerrie as a seal. He visits his wife after seven years and, putting a gold chain around the boy's neck, leads his son away to the realm of the seals. The woman later marries a hunter, but one day he brings back a gold chain and tells her of how he has killed two seals. One was old and grey and the other, a younger one, had worn a golden chain. Realising they were her loved ones, the poor woman wept with grief.

Other folk tales detail the misfortunes suffered by seal-hunters. One story tells of a man who was so convinced that his only cow died because he had killed a seal that he never hunted them again. Others are more complex, but usually the seal-hunter mends his ways after some mysterious occurrence or after an encounter with seals which have changed to human form. There is the legend of the man who, for many years, had made his living by hunting seals. He was approached by a dark stranger on horseback who told him that his master wished to do business with him. Lured to the cliff edge, the hunter was grasped by the horseman and they leapt into the sea. Under the waves, he found he was able to breathe, and was taken to the seabed where the Seal King lay wounded, surrounded by an entourage of seals of all ages. It was the hunter's own bloodstained knife that lay at the old seal's side. He was told that only the one who had caused the hurt would be able to heal it. For the first time in his life he felt ashamed of harming a seal and did his best to bandage the wound. He could only return home, however, if he promised never to hunt the seals again. This he willingly did and was duly returned to his house where he was given a huge bag of gold with which he prospered for the rest of his life.

There are also tales of seals doing good to those that treat them well, saving men from drowning or alerting other people to go and rescue fishermen in trouble at sea. The selkie would roar continuously or wail like a woman until help arrived. Whatever the story, it is clear from folklore that through history seals have held a mysterious fascination for the people sharing their environment. In a similar way to the traditions involving fairies, seals were credited with high intelligence and with other human

Sometimes early winter evenings bring a golden light to bathe the bare treetops and the castle tower

faculties, as well as having strange powers to affect the fortunes of humans.

The Columban monks are believed to have introduced the fallow deer to Kildalton around AD 900. They still thrive in the woodland and lower ground of south-east Islay. The big red deer favour the higher hill land, while the tiny roe-deer prefer the woods – although their shy natures make sightings infrequent. A drawing of the castle in 1884 shows a number of fallow on the front lawn, lying contentedly in the summer sun – just as they do now, but today the castle is a ruin inside.

Our farmhouse is only a few hundred yards away and after passing the castle nearly every day for 16 years, I have come to know it well, its own moods reflecting weather and season. At night, a faint, ghostly light emanates from soulless windows within a dark silhouette of tower, chimney and stepped gable. When the moon is bright and full, its rays penetrate the empty windows of the turret, and the wooden shutters that still remain gleam like the eyes of some nocturnal animal. When gales wrestle with the trees the wind will play noisily with any loose shutter it finds. But it is when the trees are at rest and a shutter slams shut with a bang that you wonder just how soulless the castle is.

Sometimes early winter evenings bring a golden light to bathe the bare treetops and castle tower. On long summer days, grasshoppers chirp and butterflies flit on the lawn spread with white-spotted fallow. When the castle was in its heyday the round lawn was encompassed by a drive along which horse and carriage would approach the steps that led to a large front door. Within, the hall floor was beautifully tiled and the walls wood-panelled. The wide main staircase swept up and round to balconies above. Now, the tiles are broken and the wood is weathered and rotting. Once-polished floors have collapsed, and the lattice of thin wooden struts are revealed like skeletons where the plaster has fallen from ceilings and walls. Once the largest plate-glass mirror in Europe almost completely covered one wall; now all that remains are a few broken pieces. Talbot Clifton brought hunting trophies and treasures back from his travels – a Samurai sword, caribou skins, the robe of an African witch-doctor and a jewelled box from Russia adorned the castle. The cosy library walled with oak was once full of diverse poetry and prose, volumes of wisdom and fact. Eerily intact today, the bookshelves are cold and empty and the room is dark, heavily shaded by the overgrown bushes outside.

In the hall of the servants' quarters, the summoning bells hang silent on a damp corridor wall. And down in the cold, dark cellar, one hundred empty wine bottles glint green in the light of a torch. There is no welcome here in this vacuous building. Faint whispers of the past gather to a crescendo and claim possession, rebuffing the uneasy intruder. It is a relief to leave and step back out into the reassuring daylight.

The forest used to be criss-crossed with numerous paths. In times long past there were many gardeners to tend to the policy woodland and the large kitchen garden. Between the castle and the sea, just inland from where the replica cross stands, there were the rose gardens. Planned out in squares hedged with box, they were planted with exotic plants and trees brought back from abroad by Clifton. Today, the hedges stand taller than a man, but the original layout can still be discerned. There are palm trees and bamboo, magnolia and azalea, and cultivated rhododendrons of white and red. A path leads from the castle to a spacious greenhouse, where relics of clay pots of all sizes lie amid shattered glass and rusty metal-work. Long pipes run cold around the walls, which once gave warmth to Clifton's and Violet's precious collection of orchids from faraway lands. Outside, huge wellingtonia and monkey puzzle rise majestic and proud from the rabble of wild undergrowth. And hidden in the disordered purple rhododendron lie the collapsed remains of Violet's round wooden poetry hut.

Down at the shore, paths lead to every little bay, each of a different character – some pebbled, others of silver sand. Across from the dùn, a small wooden hut with red-tiled roof peeks from the enclosing bushes and it became a tradition for visitors to carve on these walls. Some have written messages of love, others have simply signed their names with the date. All evoke the past, provoking thoughts of how the estate must have been in those days. Occasionally, people have returned to see where they wrote years ago. One man strongly believes that he had been conceived in the hut.

Following the coastal path, the next bay now holds for me a special memory. For here, beneath a cairn built lovingly at the trees' edge, the grave of little Tarzan lies – the seal pup found in Loch a' Chnuic. In the wood which stretches to the next bay, the rocks jut out here and there from the turf, some mingled with a sparkling white stone. The inlet has been fashioned to form the 'swimming-pool', one edge of which is flanked by a well-built stone jetty and a pretty bathing-hut made of quartz. The coastal path continues round the point to where the small cliffs rise and the sea lies open to the Seal Islands. Here and there man-made steps rise and fall where once walked fine ladies with their ankle-length dresses and dainty parasols. The way leads on through a shady rock cleft down to the pier below the castle and on to the bay beneath the farmhouse.

When we first came to Kildalton most of these paths were inaccessible, overgrown with rhododendron, and many were completely obscured. Some had been kept open in part by the fallow deer. Early explorations were a

The children looked like fairies as they played amongst the flowers in this magical place

delight in themselves. Struggling through the rhododendron, we discovered the hidden paths, steps and ditches. We found the broken and rusted debris discarded in the days when the castle knew life and prosperity. Undisturbed glades were still to be found, unvisited for many years but for the deer. Here the wild hyacinth and white-flowered wild garlic thrived, and in early spring snowdrops carpeted the ground. The children looked like fairies as they played amongst the flowers in this magical place. Daffodils and narcissi grew in abundance and the fuchsias hung thick with scarlet flowers. By the dùn was a tiny wooden hut; once it was thatched with heather but now it had completely disintegrated – apart from the floor made of upturned champagne bottles, half buried in overgrown turf.

In the woods, the work to open the paths has gradually progressed over the years. The replica cross stands proud and straight once more and the surrounding walks bring the pleasure of diverse views through woodland and glade to the small bays and the open ocean. Old ditches, deep with mud, have been cleared to drain flooded land. All has been done carefully by hand – hard work but necessary. For even if machines could be brought in to some areas, the soft peaty ground would be compacted and destroyed. Like a garden, the work of rehabilitation needs to be done with sensitivity.

Working in the woods and on the beaches, with the children playing

OPPOSITE:
There is no welcome here in this vacuous building. Faint whispers of the past gather to a crescendo and claim possession
(Chris Davies)

51

The replica cross stands proud and straight once more

their games of fantasy and discovery, we have communed with the nature of Kildalton and developed an intimate feeling and love for every cherished part. Taking a picnic or food to cook on a fire of rhododendron sticks, we have spent happy hours in the woods and by the shore, where passing seals surface to inquisitively watch our endeavours. The wild sea swans have learned to come for titbits and, during the spring, fluffy grey cygnets also appear riding on their backs. In some places the rhododendron branches have grown thick and tall like small trees. Huge bonfires have often continued well after nightfall when the flames give us light to work by and each ten-foot addition to the pyre sends showers of bright sparks into the darkness. In winter, the fallow would investigate where George had been working and they learned to come for any laurel he had cut, eating it even when he was still busy nearby, so used to him were they.

One such place now cleared is the site of the old Ardimersay Cottage, lying between Kildalton Castle and the sea. The area, once completely enclosed by rhododendron, is below a high ridge and a wide path leads from the castle down a steep slope to the wooded glade below. Where the building stood, tall spindly sycamore compete for light, and in the spring the ground is carpeted with profuse wild garlic, the pungent aroma liberated underfoot. Built into the side of the cliff is a cold stone hut, known to have been the larder for the castle. The children used this as a playhouse while we worked. The glade leads to the bay with the bathing-hut and, part way down, covered in moss and small ferns, the tiny domed house of a well lies. Only a few feet high, it is built with small stones and has a green wooden door. We call it the Fairy Well and it was so well hidden that only those who knew its location could find it.

Ardimersay Cottage was a regency pavilion, with latticed windows and thatched with reeds. It was built for the Campbells who were the Lairds of Islay until the mid-nineteenth century. A portrait of Lady Eleanor, wife of Walter Frederick Campbell of Shawfield and Islay, was painted in the cottage. She stands by an open window with the view looking across woodland and sea to the Ardmore hills across the bay.

The wild sea swans have learned to come for titbits

In 1855, John Ramsay bought the estate. He was a Member of Parliament and a distiller who prospered by exporting whisky to the United States. Although he was very fond of his home at Ardimersay Cottage, after attempting to make it more weatherproof by replacing the reeds with slates, he eventually decided to move. In 1870, Kildalton House, now known as Kildalton Castle, was created on the higher ground above the cottage. It is thought that it took up to six stonemasons five years to build this baronial mansion with Bath stone. It is now a listed building. For some time the cottage was still used as a smoking-room before finally being demolished. Most of the land of An Lanndaidh is now protected as a Special Site of Scientific Interest in a Regional Scenic Area and the importance of the policy woodlands of the castle grounds, now regarded as an arboretum, has been recognised.

John Ramsay, his second wife Lucy, and one of his three daughters are buried together overlooking the sea. The three stone crosses stand a little above the coastal path that follows a cliffy ridge to the pier. A narrow but deep cave cuts into the rock face, where pottery, stone objects and animal bones from the Bronze Age have been discovered. The beautiful bay below is called Port Luing Mhic Ruaridh at the head of which is our farmhouse built high above the beach. This sheltered inlet has probably been used over many centuries as a harbour and may well have been named in the fourteenth century after Amie MacRuairi. She was the first wife of John of Islay, one of the MacDonald Lords of the Isles.

For many years, although the seals would often swim past this bay, they were never seen to haul out at the little islands in the centre. It was only after I began to rear seal pups and bring them to the shallow waters by the beach that the seals came to lie on these rocks, perhaps for the first time in hundreds of years. I can see them from the house, but often take a walk down to the pier for a closer look. Concealed by rocks, bushes and trees lining the coastal path, I am careful not to disturb them. Sometimes I recognise one that I have released myself which gives me great pleasure. But often, now, wild seals rest here lazily in the sun – the same ones returning, to whom I give names.

I wander alone under the trees, their solid limbs soft and green with lichen and moss and fern. Content in my solitude, enjoying each spectacular view, with no one to listen save the ever-curious seals, I can sit at the sea-edge singing and composing my songs. The ocean and the sky change with the weather and the trees and flowers alter with the seasons. People come and make their transitory mark upon the land which, in return, claims their deep affection. And the essence of Kildalton's beauty endures through time.

INTO THE NIGHT

Night, when the forest is black
I will stay by my fireside.
Night, when the full moon is high
I will hide from the stillness.

You come alive when the shadows fall,
you stay away from the lighted hall.

The beautiful bay below is called Port Luing Mhic Ruaridh

And over the mountains the sunrise will bring the light.
And over the water the first bird calls in flight,
'It's a new day; it's a new day; it's a new day.'

You fall asleep in the secret place,
one day I will find you and know your face.
And over the mountains the last rays fade into the night.
And over the water the last bird calls in flight,
'It's a new day; it's a new day; it's a new day.'

FOUR

Melody in Green

Melody in green,
springtime.
Lazing by a stream,
love time.

I will follow you
a-dancing through the willows.
Feel the summer grow,
the sun shine.
Riding my pony through shady woodlands.

Melody in green,
I want to sing your song,
lie in your meadows,
and talk to the trees.
How I love you, blossom tree.
Do you know, I really feel again.
Feel again,
feel I'm me.
Melody.
Melody.

Melody in green,
you're a simple song.
Lazing by a stream,
here I belong.

Holding on to my memories,
to last me the long winter.

OPPOSITE:
Only to rise again
elsewhere, with eyes
still fixed on me

Feel the spring again,
the sun shine.
Riding my pony through shady woodlands.

Melody in green,
I want to sing your song,
lie in your meadows
and talk to the trees.
How I love you blossom tree.
Do you know, I really feel again.
Feel again,
feel I'm me.
Feel again,
feel I'm me.
Melody.
Melody.
Melody in green,
melody in green.

As Jason broke into a canter across the soft grey sand, I regarded the gleaming waters in anticipation. But for a couple of eider duck bobbing on distant gentle waves, the bay was empty. I had ridden from the farmhouse, past the fields and the kitchen garden, down to the southern end of the peninsula. Leaving the beach, we trotted along the muddy path and made our way through yellow gorse bushes to the edge of a rocky shore where I dismounted and allowed Jason to graze on the thick, coarse grass. Here the water was deep, and I knew that if the seals were near, they would come.

Even when still in England, I would sing while riding Jason through the New Forest. If composing, it would, on occasion, dawn on me that the rhythm of the song was echoing the sound of his hoofbeats. As I sang by the shore at Seal Bay, he carried on grazing regardless – just as he would take no notice of the seals that were coming into view. One by one they gathered from the placid swell, surfacing from the secret depths, their unhurried advance serene as the soft melody of my song. When close enough, I picked out familiar faces – the old bull seal who had frequented these shores the winter through, and the young ones I had watched chasing each other in leaps across the waves. Together they stopped and stared till one would sink beneath the inky waters, only to rise again elsewhere, with eyes still fixed upon me.

As I mounted again and Jason picked his way over the rocky ground, the seals followed us along the shore till the sea became shallow and we withdrew to our own worlds once more.

This end of the peninsula rises gently from the shore through salt marsh and wet-heath to thick scrub of oak, hazel and birch. The grasslands have, in the past, been drained with wide ditches running parallel down to the sea. Now, they lie filled with mud and vegetation, scarcely discernible from the surrounding land – a testimony to the efforts of past generations to gain control over the water-logged ground, and to nature's ability to reclaim it. Jason, a horse of irrepressible character, loves to jump across these, as he does the many deep holes pitting the short green turf at the sea edge. When the tide is well out between the peninsula and the island, the large expanse of exposed sand is perfect for a good gallop. A long line of stepping-stones traverse the breadth of the gap – evidence of a bygone age when prosperity made such a labour-intensive project possible.

I just love to sit and watch them from a distance, getting to know the ones that frequent a certain site

The low-lying land is interspersed with rocky outcrops and, further inland, long lush glades alternate with high-cliffed ridges crowned in dense thicket. In spring, the pure notes of the cuckoo sound over a sublime landscape. Eider, mallard and shelduck hide their snug nests in the heath and marshland. On a grassy rock outcrop just off the main island, a pair of mute swans return each year to breed. Then, with the bloom of summer, butterflies and moths vacillate over the pale-pink orchids, purple heather and vetches that colour an abundant sward. Sometimes a shag stands alone, motionless as a statue on a sea-bound rock; dark wings spread out like a cloak to dry. Oyster-catchers strut about at the sea-edge searching for food with their sharp orange beaks, and rise, disturbed, with a piercing cry.

The main island of Imersay is usually dangerously sodden in winter, but with summer's heat Jason can happily negotiate the drier heath. Seals frequent the haul-outs on the many surrounding islets and skerries and take little heed of a horse passing on the island. Walking from the house up the farm fields and through pine woods towards this end of the peninsula, from the higher ground I am able to see, with the use of binoculars, whether or not the seals are out on the rocks. Through the years, I have learned to judge by the weather and the time of year where seals are likely to be found and what they will be doing; but their ways are also governed by variables of wind direction and state of tide. It is a good walk all the way down to the sea, across to the island and out to where I can observe the seal haul-outs. I love to just sit and watch them from a distance, getting to know the ones that frequent a certain site.

There was one summer when I often visited one area by Imersay Island because of the presence of a particular female grey seal. She had chosen to join a group of commons of which there were usually ten. The piles of craggy rocks offshore here are arranged in such a way so as to form a lagoon against the island, sheltered from the rougher sea outside. The seals would be lying on the lower slabs, the grey amongst them. Her beautiful creamy-white coat was blotched with dark-grey patches and her demeanour was that of a queen surrounded by her entourage. Sometimes she would be in the water; so restful to watch when compared with the commons, who never seem to be content with staying still for long. Queenie would be in the middle of the lagoon, head and neck above the quiet water, gently turning her attention to one side or the other or, if I was singing, she would stare straight at me with her enormous eyes in a steadfast, intent gaze. Some of the seals were quite happy to stay lazing on the rocks – even at that short distance, they were not afraid – while others would come over to me.

Another time, not far from this spot, I was singing when the tide was high and a fair wind was driving the waves inshore. About twenty seals were in the water, their haul-outs mainly covered now by the sea. Together they swam towards the head of the inlet where I sat and slowly dived in front of me. Then more would come and they dived in turn. I realised with amusement that, as if playing a game, they were repeatedly swimming underwater to the back of the queue and coming inshore with the wind behind them and the choppy tide carrying them forward.

One sunny autumn afternoon, I was playing the violin on the other side of Imersay Island. Sitting on a low rock in the shallow waters between the island and the peninsula, I became aware of the snuffling and snorting of a seal nearby. At first I could not see him and carried on playing, but, eventually, a seal appeared round the side of a nearby rock and eyed me as he glided along. I was amazed at how close in he had come in quite shallow water and, emboldened by his forward behaviour, I walked waist-deep a few yards into the sea, still playing the violin. I was facing the peninsula and, beyond a copse of dark pinewood, the verdant dome of the Fairy Hill rose against a pure blue sky. I was spellbound by the beautiful scenery and enchanted by the close presence of this beautiful wild animal.

He was a juvenile and his bright-eyed, whiskered face showed all the innocent inquisitiveness of youth. He was soon joined by another of similar size and in the flat, calm water I played and watched them, enthralled as they plied slowly back and forth on the same level as my violin. They were about ten yards from the shore where the water was deeper. Now I was experiencing the company of wild seals from a new angle, having entered their domain. Eventually, they wandered further and further away on their dives, but it was not from fear; merely, it seemed, they had other things to do . . .

George was baptised in the waters of Seal Bay. As friends and members of Port Ellen and Bowmore Baptist Churches gathered on the shore, George was aware of two seals watching from a distance. His audience lingered as George, wearing his exclusive Bens of Jura T-shirt, was totally immersed in the cold water. George and I had been married at this same church in Port Ellen, and my father and I arrived at the Baptist Church in a gig drawn by Emma's grey pony, Misty.

Born in the New Forest, he had been saved from the European meat market at three years old. After moving to Islay, we became interested in breaking him to harness and were introduced to a man who was to become a dear family friend – Sandy McCormick. Sandy loved horses – his father had farmed at Keills on Islay where Sandy worked with the Clydesdales at ploughing and harvesting; carting draff from the distillery maltings for cattle feed and sheaved oats to the mill. He was delighted to help with Misty, who learnt to pull the gig and the cart. Eventually, we bought a young chestnut Clydesdale gelding called Prince. There had been no Clydesdales on Islay for many years and it was a pleasure for Sandy to be able to work with one again.

It was not long before Prince had a new name, however. One very hot day, two of my stepchildren, Sarah and George, who were about seven and five at the time, had been playing in the courtyard of the farmhouse. Suddenly, in great excitement, with wide-eyed worried faces, they came running over to me. 'Big horse is dead, big horse is dead!' shouted Sarah.

'Biggles is dead, Biggles is dead!' echoed George.

Upon investigation I discovered the Clydesdale flat out and blissfully asleep in the comfortable shade of our large chicken shed. Being a great fan of the intrepid pilot, Biggles, I allowed the new name to stay.

Sandy ploughed the old kitchen garden with Misty and Biggles working together. He would be up early in the morning getting them ready. Their muddy coats were groomed and he tied up the Clydesdale's tail in the old way, for, as far as Sandy was concerned, everything had to be done right and he took great pride in all that he did with the horses. He brought manure to the garden in the cart to spread on the neatly ploughed earth. The chain harrow prepared the ground for barley seed and the drill plough made high ridges in which to grow potatoes. A large area of turnips was grown for the horses and a small plot ploughed for a variety of vegetables.

Whatever task was in hand, it was always satisfying to see the Clydesdale and the stocky pony working side by side, up and down on the fresh-turned earth, or to hear the rumble of the empty cart on the drive, heralding the return of Biggles for his well-earned evening feed.

The garden still has old apple, pear and cherry trees from the days when it was used to supply the castle with all the fruit and vegetables it needed. Large, brick cold frames lie at the sunniest end of the enclosure and, at the other end, the garden is well sheltered from fierce sea winds by a high stone wall. In spring, an abundance of daffodils flower in a central copse and the fruit trees blossom in a profusion of white and pink flowers.

In one corner of the garden there is a large wooden shed where tools and flowerpots were once stored. Late one spring, a sparrow-hawk attacked a family of swallows which was nesting in the shed, causing the tiny birds, still in their nest, to fall to the ground. The mother had such a fright that she flew away, and when it was obvious there was no hope of her returning,

When they were a few weeks old and their feathers well grown, I helped them learn to fly

I took the four babies home to rear myself. I kept them in a snug box by my bed with the heat of a lamp overhead. Here they were handy to feed, but I had to find a way of catching flies for them. On sunny days, I would find a place, usually a smelly one such as the midden, where flies are abundant. They would land momentarily on my body and I became expert at fly-swatting, popping each new morsel into my jamjar. I fed the little birds, who at first had no feathers, with a pair of tweezers, discovering that if I passed my hand over the light, imitating the parent bird's shade on entering the nest, all the tiny beaks would open with a loud, demanding 'cheep, cheep, cheep'. They were called Eeny, Meeny, Miny and Mo – Eeny being the largest and the rest named in descending order of size.

When they were a few weeks old and their feathers were well grown, I helped them learn to fly in the bedroom with the curtains drawn. At first, they would fly from my hand on to the bed and then their flights became increasingly longer and more adventurous. Everything seemed to be going well – but one by one they died, until only Eeny was left, who continued to grow bigger and stronger. Every summer the castle and the farmhouse outbuildings are alive with the activity of swallows. Every shed seems to have at least one family, and the fledglings fly around and perch in lines on the rafters. I knew that the time must be nearing when my baby would have to learn to catch his own food. One day I was outside with Eeny sitting on my arm, when a dozen chirping swallows came swooping over my head. Suddenly, Eeny was off with them, and they all went flying gracefully to the back of the buildings and perched high up in a tree. I left his box on the ground nearby and kept an eye on him; but, eventually, he disappeared from his branch and I never saw him again. I hoped he would make it to Africa in the migration and return to Kildalton after the winter. Each spring, as the swallows gather about the farm buildings, I think about Eeny and wonder if he is among them.

Another youngster which I took in was a male blackbird, who was found alone in the woodland drive. He was left until nightfall, and when he was still there on his own in the dark George decided he would have to be looked after. Tweety Pie progressed well and became part of the family, learning to hop about under the kitchen table picking up any crumbs the children had left. The only real problem arose when it was time for him to make his own way in the world. He certainly had no intention of leaving himself. I would take him down to the garden, which is about half a mile from the house, where he hopped about eating any worms or grubs that were uncovered by my digging. The first few times he followed me all the way back to the house – however carefully I tried to sneak away. But one day he stayed in the garden and the ties of his unusual upbringing were broken. Some time afterwards I noticed that a beautiful mature blackbird was frequenting the shrub at one side of the potato plot, and was convinced that my Tweety Pie had made it his home.

One early summer evening George was driving through the woods near Seal Bay when his headlights picked up a small, feathery bundle in the road. He stopped the car and got out to investigate, discovering that it was a very young tawny owl, which he thought was dead. The downy bundle was still warm, however, and realising the bird had been stunned by a car but was still alive, he quickly brought it home. We gave it warmth and shelter in an open box in the living-room and hoped for the best. After about an hour it showed real signs of life, standing up in the box, but was still dopey and wobbling. We were worried about it because, although there was no blood, we did not know how severe the head injury was. A little later our fears evaporated when the little tawny owl flapped his russet-brown wings and flew strongly up to the top of a high cupboard. His swivel head turned toward us as he took in his new surroundings with beautiful dark eyes. We decided to keep him until the next evening to give him time to recover completely, and gave him the freedom of the room for the night.

As darkness fell the following day, we put him in a box and drove back to where he had been found. It was a still and starlit night as we took the owl a little way into the woods, not knowing how he might react or if he would be safe once we let him go. George opened the box and we waited. I felt he was getting his bearings, slowly realising that he was back in his own woodland. Suddenly, making a soft call, he ascended into the darkness of the trees and, like music to our ears, we heard from the thick treetops the louder cry of the mother echoing his.

There are also barn owls in the woods and I like to hear their shrieking after dark, occasionally seeing one in flight during the day with its striking white plumage. Clifton would lie awake at night enjoying their screeching in the trees and around the castle – although some find their eerie call disquieting, and through history they have had a reputation as a bird of ill omen. The Cliftons, however, have a tradition that the owl brings good fortune to their family.

Islay has an unusual ecosystem, having no foxes, badgers, squirrels or moles. Until they were introduced in recent times there were no hedgehogs either. But the island has its own sub-species of common vole and field vole (*microtus agrestis Fiona*) and is home to certain animals and birds that are becoming rare elsewhere. Otters thrive in many coastal parts of Islay and Jura, and Kildalton is no exception. From a distance, the otter's head could be mistaken for a small seal when in the sea, but their movements are usually quicker, and when they dive they can be identified by the slim, lithe curve of the body and an appearance of the long tail. I have often seen them running across roads, fishing in the sea and darting about exploring on rocks and islets. In the woods, their short, loud bark can sometimes be heard as they call to each other. Whatever the activity, they always seem to be full of irrepressible energy. I once found one dead when exploring among trees near the castle, but I had no reason to believe that it had died from anything but natural causes. Although sad at her demise, I was fascinated by the thick, tapered tail, webbed feet and broad, flat head.

My closest and most privileged encounter with an otter was at Seal Bay when on my way round the shore to visit some seals. I was just passing a large rock between myself and the sea when he caught my eye, just a few yards offshore, busily diving and swimming about in one spot. He had no idea I was there quietly watching him from behind the perfect natural hide. I observed his moustache of stiff whiskers, bright little eyes and tiny ears set wide apart. I felt sure he was hunting but he never came up with a fish and, after a while, he swam off. My daughter, Hannah, was with me and although she was only three years old at the time, she had learnt that when mummy or daddy made signals to be quiet when out walking there was a special bird or animal nearby.

His head and tail were stretched high in the air and, as he posed, he looked as
though he was performing some yoga

OPPOSITE:
She would love to sit
on the Fairy Well and
make a wish that when
she was older she
would play to real seals
like mummy (Chris
Davies)

Hannah was born on Midsummer's Eve in 1987 and she was also given
the name Titania after the Fairy Queen in Shakespeare's *A Midsummer Night's
Dream*. When she was old enough to hold a little violin of her own accord
she would line up all her fluffy toy seals and play to them. She would love
to sit on the Fairy Well and make a wish that when she was older she would

play to real seals like mummy. Hannah had to have her own pair of binoculars and I was always careful that she had them with her on our expeditions – especially after the time when I gave in to her pleas, letting her look through my good pair, and she dropped them in the sea.

When I was becoming heavily pregnant with Fifi we had to finally give up our walks down the peninsula to the islands, for as well as negotiating bogs and rocky places myself I had to help Hannah over the difficult terrain. But I could not give up being with the seals, and nearly every day Sandy would drive us along the narrow public road that runs past Seal Bay, where I would observe them with my binoculars while Hannah happily drew all over my notebook. There is a string of rocky isles quite close to the road and another further out near Imersay Island, and these are often laden with common seals. Visitors stop on the road to see them and this is where I saw my first wild seal all those years ago. His head and tail were stretched high in the air and, as he posed, he looked as though he was performing some yoga.

In the months when the sea is swollen and rough with winter gales; the few rocks that are still exposed lie empty. Sometimes, in sheltered water, a seal can be seen bottling – resting far below the waves and returning to the surface for air every ten minutes or so. Only his head will appear as he breathes for less than half a minute before gently slipping beneath the water again. They always seem to be still dopey from sleep while doing this, completely ignoring me if I happen to be singing. While at the surface his heart-rate would be 120 beats per minute, falling to around 40 while underwater, and his blood is specially adapted to store large amounts of oxygen during his sleep beneath the waves.

In better weather, when more of the rocks are above water, the few seals that have remained faithful to this haul-out gather in the sheltered inner rocks, joined by passing strangers – the more exposed far reef only being used on particularly fine days. There are usually only about eight that stay in this particular area all winter, rising to up to twenty with the arrival of incomers. But in early spring, with the occurrence of a calm and sunny day, the rocks are suddenly heaped with shining bodies of all sizes. On one day in April, I noted a record total of 64 seals, when it seemed that every available rock was occupied. Among these were ten pups born the previous June. A few seals were swimming and now and then one would porpoise high above the flat water. Others were still searching for space on the islets, hauling out dark and glistening and shuffling about until comfortable. Those that had been lying on the rocks a while had dried off and their different markings could be seen.

Sandy told me that years ago there was an old man living in the only nearby cottage at Maol Bhuidhe who had names for all the seals and could point out each individual. I, too, have names for seals that frequent these rocks, and enjoy noting unusual markings on passing visitors. There was one large bull seal who I named Elly. Although he was always on the far rocks, I hardly needed the binoculars to identify him as his enormous size gave him away. He was usually joined by three white seals with dark flippers, tails and noses who I called 'the three pandas'. One distinctive seal had similar markings but his pale body was very spotty. One (who I named Pepper) was grey with speckles all over. Yet another had a spotted-grey head but his body was of a lighter colour and his tail and flippers were dark.

Sometimes I would notice a seal that did not seem well, particularly if it was having difficulty in diving. I knew that was likely to be caused by a bad dose of lungworm, but there was nothing I could do to help. Even in laters years when I had some expertise, I would only be able to come to their assistance once they had become so ill that they were weak and could be easily caught somewhere out of the water. Others had injuries – bull seals in particular were prone to neck wounds caused by fighting during the mating season – but in the main nature would heal these. I also sometimes saw seals with eye injuries and there was one that I felt sure was blind in one eye. It is said that a completely blind seal can manage to live and feed itself, fishing with the sensitivity of his whiskers.

There was one incident with a young seal that still saddens me, for it could have been prevented if only people realised the suffering that they can unknowingly cause. The sea is used as a rubbish dump and the evidence of this is washed up on the shores of Seal Bay every day. The seal was spotted lying on a rock in the normal way but its neck was encircled by a piece of tough plastic strapping. As it was already growing weak, various attempts were made to rescue it from certain death since, as the seal grew, the plastic gradually choked it. All our efforts failed and we called on British Divers' Marine Life Rescue. But before they could arrive the seal disappeared from its usual haunts. I have heard of similar incidents from other people interested in animal welfare.

The beauty of the coastline is marred by thousands of plastic objects. Detergent bottles are washed in on the tide along with fish-netting and all sorts of discarded packaging which become entangled in the seaweed. Plastic bags of different colours and sizes are blown into the bushes and trees at the sea edge. Rubber tyres and plastic boxes sink into the soft sand of the beach. Heavier items get stuck before they reach the shallows, their carcasses evident at low tide – an armchair, oil drums, even a fridge. The harmonious mixture of greens in the waterside foliage and the brilliant yellow of the wild iris-bed are dotted with a multitude of bits of glaring white litter.

With friends and family we work hard at beach-cleaning, spending several days at a time in one spot to clear it well. But we know that with the next tide the relentless onslaught from the outside world will begin again; an eyesore and a hazard to wildlife. Odd bits and pieces with Russian and Oriental writing turn up amongst the debris, but, more ominously, a yellow wooden triangle, printed with the radiation warning sign of a skull and crossbones, was found just along the shore at Ardbeg. It had been placed in the sea at the outflow of the Windscale nuclear reprocessing-plant in north-west England by Greenpeace as an experiment to indicate where the tides carry the effluent. Windscale is now called Sellafield but the currents and the tides have not changed.

In May, the expectant seal mothers are looking very fat. As well as carrying the pups, they have a lot of blubber built up over the winter which will help them through the suckling period when they cannot go on long fishing trips. In June, they give birth in the sea by the big islands beyond the Fairy Hill. But there are still a fair number left in Seal Bay – juveniles and females too young to breed, as well as adult males. In July and August, the seals go through their annual moult when the coat looks dull, brown and patchy and individual identification becomes difficult.

The seals spend more time hauled out, lazing about in the sun, which helps to speed up the activity of the skin cells. Because of the mating season, the male seals' behaviour in July can be fascinating to watch. I have seen a seal crossing from one end of the bay to the other, rolling on its side and slapping the water with his fore flipper at intervals with a resounding 'thwack'. They often play with the seaweed at this time, holding long trails of it in their mouths and waving them about. I was playing the violin once, with George watching from a distance, and I had my eyes on all the seals in front of me. He told me afterwards that to one side, out of my direct sight, there had been a large seal flinging the kelp about wildly. I was pleased that the bull seal had felt so relaxed in my presence that he carried on with his normal activities. What other wild animal would do so?

I have always enjoyed playing the violin outdoors and have taken my instrument with me on holidays abroad. In Switzerland, I played while travelling by train through the Alps, and in Austria on a remote glacier as the sun rose over the mountains. I remember my fingers being so cold that

I could hardly feel them on the strings, and there have been days when they have been just as numb when playing to the seals at home. In Canada, the violin came with me as George and I walked in the glorious Rockies. On one occasion I was playing by a beautiful turquoise lake when a porcupine climbed halfway up a nearby pine tree and, clinging to the trunk, stared at me with something like disdain. If he could, I think he would have reported me to the Park Ranger for cruelty. We came across a black bear one day and George began to talk to it in what he thought was a friendly manner, pitching his voice somewhat higher than usual. The bear squealed like a cat and charged towards us! He was so quick he could have been upon us in seconds, but he stopped dead in his tracks as suddenly as he had started and sat regarding us steadily. We beat a cautious retreat and I decided not to try playing the violin to any bears.

At Seal Bay on a clear night, the silver radiance of the full moon reveals glistening rotund bodies reshaping the ancient black rocks. A lustrous beam crosses the island, the skerries and the water, and the sea laps gently in the darkness of the shore. Standing in the shallows, statuesque on one slender leg, a heron is caught in the incandescent moonlight. There is a splash and a seal is leaping and leaping again – a black torpedo curved against the gleam in the waves. There is a cloud that passes across the path of the moon, and the light is no more, and the rocks are bare rocks, and the waters of the sea are in darkness.

DAYS GONE BY

The birds can sing, the sun can shine,
but nothing moves your heart.
A butterfly, you pass him by
without a loving thought.

I remember you in the woodland walking,
I can see you talking to the trees.
We watched the catkins in the breeze,
for you always loved such things as these
in days gone by,
in days gone by.

'Though I take your hand and smile at you,
your thoughts are far away.
If you could only turn around
and laugh your tears away.

I dream of you with your smiling face on,
just a memory from the days gone by.
If you could feel the beauty round you now,
I'd surely see the smile from days gone by,
from days gone by.

In early spring the snowdrops came
in clusters on the ground.
You never even noticed them,
I saw you tread them down.

I remember you in the woodland walking,
you used to show me tiny buds come through.
If you could feel the beauty round you now,
I'd surely see the smile from days gone by,
from days gone by.

FIVE

Today the Seals

I pushed my way through a green forest of thick bracken. It was a hindrance to me now. The violin case was heavy in my hand and awkward as I tried to run down the open glade in my rubber boots. With annoyance I realised that in my haste I had forgotten the binoculars, but I did not want to turn back, for I had to get down to the island.

Below me the peninsula basked peacefully in the last warmth of a late summer sun, the dark blue sea a shimmering expanse ebbing to a pale horizon. Water crept round the stepping-stones to the island as the tide made its inexorable progress. Hurriedly, I splashed through the puddles, slippery with weed and, making my way round the muddy shore, at last arrived breathless at the lagoon.

A seagull filled the quiet air with its raucous warning cry and I saw a drowsy seal lift his head as I scrambled over the pile of boulders to the water's edge. Bone dry after a long day in the hot sun, a group of seals lay sleepily on the higher rocks, some still dull brown in their moult. Those lower down were nearly awash and dark with the incoming tide. They stretched up with graceful necks and curved hind-flippers, peering at me with dozy eyes as I approached. A couple were already in the water, and a V-wake formed behind each head as they glided toward me. They seemed to be swimming in the normal way, the body hidden beneath the surface. Some of the seals on the rocks were blinking and had a sticky-looking discharge around their eyes. Was it more than usual?

I began to sing but there was tension in my voice and tears were in my eyes. A few more seals slipped off the rocks joining the others, but the words of my songs were meaningless for there was no joy in my heart. In the silence we regarded each other, their big dark eyes in puzzlement and mine in sorrow and confusion. I picked up my violin and played 'Islay Mist' in an effort to recapture something of the serenity there had always been in our fellowship. The melody flowed like a balm on my emotions, as the seals swam up and down; so beautiful and innocent, wild and free.

A cool breeze began to ruffle the calm sea as the last seals slid into the water. A breath of wind blew the hair across my face as the unstoppable advance of the tide from far-off shores brought the outside world to mind. We were a blessed enclave on a sick planet; but the ones I loved were in terrible danger and I had no way of defending or warning them.

A new song came to me, with a rhythm that ebbed and flowed like waves on the shore. Love, entwined with sorrow, bore the sad sweetness of the melody. Anxiety and guilt yielded words of forewarning and anger. Anxiety because I knew that a death virus was rapidly spreading through seal populations; anger that there was no help offered from the government to nurse sick seals; and guilt at belonging to the human race that polluted the oceans.

I started as the piercing cry of a seabird racing low over a nearby shore wrenched me back from the depths of creative meditation. Night was falling fast and I suddenly realised how cold it had become and that the seals had wandered further away to fish. I made my way home, stumbling over tussocks and rocks in the half-light and wading through the deep water that now separated the island from the peninsula. What had I gained, when today the seals were well but tomorrow or the next day the virus could suddenly hit them? All I had were the beginnings of a song forming in my mind. I walked through the dark pinewood, in an eerie stillness broken only by the shrieking of owls and, as I entered the open fields, a dull, lifeless sky was intermittently pierced by the bright, sweeping beam of the lighthouse on Churn Island. The farm buildings were hushed now, the swallows long gone to bed, but with the fall of darkness the bats had come out from their roosts in the sheds and were fluttering busily in the dark; hunting on the wing like swallows of the night.

I changed out of my wet clothes and, seeking solitude, went to the studio. Here I could work on the song. I had all my old recording equipment, suitable for rough demos, that I had brought up from England, but it was nothing compared to the 16-track system that George had decided to buy me in June. Whatever prompted his decision at that time will always remain a mystery, for I had already been on Islay for 11 years with nothing more than a vague notion that, sometime, I ought to do something with my songs.

In the days that followed I visited the seals whenever I could and each time was relieved to find them healthy as they sunbathed and cavorted in the sea, blissfully unaware of the danger they were in. I sang my new song to them as it unfolded, and played them the violin parts that I had

Big dark eyes in puzzlement

composed. But a feeling of utter helplessness pervaded these days of uncertainty. The terrible television pictures of thousands of dead and dying seals struck fear into my heart. In some areas over 50 per cent of the population was being decimated. Of course, I hated to see the suffering and destruction in other places, but I prayed especially for the seals I sang to on Islay. I hoped they would escape the virus and that the relative purity of these waters meant that they were stronger and more able to resist the disease. I also felt that the longer it took to arrive here, the more chance this colony had of avoiding an epidemic, as once the moulting season was over there would be less seals hauling out together to spread the virulent disease.

There were no seals to be seen on stormy days when the tide was high and only the very tops of haul-out rocks were exposed. My anxiety would increase and I longed for the calm weather to return, bringing them back to the little isles where I could keep an eye on them and know they were safe. If any were sick, the gales whipping up powerful ocean waves would finish them off before they could be found and given help. The suffering of the seals was highlighting the fact that mankind is polluting the seas, using the world's beautiful oceans as rubbish dumps and sewers. If there was something badly wrong with the health of animals at the top of the food chain – and pollution is strongly indicated as the cause – we have to consider seriously the implications for the marine environment as a whole and, ultimately, for ourselves.

TODAY THE SEALS

Dear seals, I've often watched you,
you're part of the islands I love.
I've sung you songs of many things
as the seabirds ride the changing winds above.
You come to see what's going on and stay with me a while,
I've often thought that if you could, you'd wave to me and smile.

Yes, you're part of these island shores,
like the curlew and the yellow gorse.
I've loved you like a part of me –
wild things in the restless sea;
wild things in the restless sea.

So far we seemed from sorrows
that trouble wildlife in the world today,
but now the tide is turning,
my hopes for you are fading away.
Feeling guilt and helplessness, is there nothing I can do?
Is this the last time I can play my violin to you?

There's a cloud over this ocean sky,
how many of you have got to die?
Today the seals, tomorrow who'll
suffer while mankind plays the fool?
Suffer while mankind plays the fool.

I thought you'd always be there,
now I'm so fearful for you.
There's danger in these waters –
is there nothing anybody can do?
Today you're bathing in the sun and playing in the sea,
tomorrow island birds may sing a requiem with me.

The rocky isles across the bay
lie empty on this stormy day.
This time have the seals gone
where they can't hear my farewell song?
Where they can't hear my farewell song.

There's a cloud over this ocean sky,
how many of you have got to die?
Today the seals, tomorrow who'll
suffer while mankind plays the fool?
Suffer while mankind plays the fool.

For he was of an innocent race that peopled the sea, and I belonged to a society
responsible for gradually destroying his world

From the house I had watched the storm unleash its sombre fury on the Seal Islands. Dark waves broke with white spume on bleak skerries beneath a mournful sky. The wind abated but left the troubled water aroused and restless as it beat its slow, steady drum against the shore. For me, the sea had become a familiar companion, but now it was always on my mind and I watched the seals closely.

As the storm calmed, I fetched my violin and wandered behind the castle on the off-chance that a few seals might be swimming again by the dùn. The bay is sheltered and would be one of the first places to feel the slackening of the wind. From the top of the dùn I played 'Today the Seals', looking up to Clifton's grave on the Fairy Hill, then turning towards the Seal Islands, hazy on the clouded horizon. Something long and white caught my eye on the pebbled shore of the wooded bay below. My mind was in a whirl as I rushed down to the beach. It was, indeed, the small body of a dead seal. As I knelt at his side, wild emotions flooded through me. I had never been so close to a seal before, yet never so far apart. For he was of an innocent race that peopled the sea, and I belonged to a society responsible for gradually destroying his world.

His furred skin had been torn off as he was mercilessly tossed about on the rocks. Through my tears I noticed how hand-like his fore-flipper was that lay limply at his side. I would never see his beautiful eyes again for they were closed now, forever. The moment was even more poignant for I had first played to the seals from the dùn. Now, as if a trust had been broken, a young dead seal had been washed up here, and I knew the disease had arrived.

I was desperate to do something to help the seals and was heartened to find that others were too. The Islay and Jura Seal Action Group was formed among concerned local people. Dr Malcolm Ogilvie, who has worked with wildlife on Islay for many years, was voted in as chairman, with George as vice-chairman and Bank of Scotland branch-manager Aeneas Nicolson as treasurer. Aeneas is well known locally for his love and knowledge of otters. Diver Don Bowness and his wife, Dot, who regularly fish for clams off Kildalton, are very fond of the seals and, knowing the coastline and waters where the seals live so well, they agreed to help in looking after and piloting a boat.

The general aims of the group were: the welfare of commons and greys on Islay and Jura; to monitor and research the seals in the long term; and to raise funds to meet these aims. Other members included the local veterinary surgeon, Dr Janet Berry, Janette Young from the Nature Conservancy Council, Mike Peacock from the RSPB and Dr Eric Bignal who has much experience in wildlife and nature conservation issues. Further valued support came from local councillor Mr McKerell (although he was unable to attend meetings), and a representative from Jura, David Mack. Initial discussions revolved around the immediate need for more information on caring for sick seals and the requirements of a sanctuary. We also needed a boat that was able to go out safely into the dangerous waters off Islay in inclement weather conditions.

Over the weeks that followed, dead seals were discovered up and down the south-east coasts of Islay and Jura. There was a young one found on the wide sandy beach at Seal Bay where I ride Jason. Since it had not been long dead, we hoped to obtain a blood sample – possibly within 24 hours of death. We were issued with kits from the Sea Mammal Research Unit to take these and other samples from dead seals.

It was a frosty November evening when I went down to the bay with George Jackson of the Nature Conservancy Council. His daughter is married to marine mammal researcher and author of a book on common seals, Dr Paul Thompson. It turned out that the seal had been dead too long for the blood sample to be of use, but we took part of the jaw and a section of blubber, measured its length and girth and confirmed its sex. I was certain this was one of the young seals I had known well from playing the violin on the shore at Seal Bay, for his markings were familiar. I found the whole operation quite distressing but steeled myself in the belief that, ultimately, any research would be to the benefit of the surviving seals. There we were in the freezing cold with protective rubber gloves on, cutting up a lifeless, pathetic body. He had been cast up like the rubbish on the beach we had been clearing; just another piece of debris from the foolish waste of man. Another two were found on the shore on the tidal island just a few rocks away from where, on hearing of the disease, I had begun to compose 'Today the Seals'.

In the summer of 1988, the previously unknown phocine distemper virus hit seal populations in the North and Baltic seas. Scandinavia, Germany and the Netherlands were affected before the disease crossed the North Sea to the Wash in August, and then spread to Scotland and the Irish Sea. It was mainly the common seals that were being devastated in the epidemic and, once an animal showed symptoms of the disease, death usually followed within days.

With a discharge from both nose and sore, inflamed eyes, the sick seals coughed and sneezed and had difficulty breathing. They were lethargic and had problems with their nervous system, including an uncontrollable trembling and shaking. The digestive system was also affected and enteritis caused diarrhoea. Because of the strangely arched back which the disease produced, and their difficulty in diving, seals with the disease could be detected in the water.

In April, an unusually high number of seal pups had been found dead and dying in the Kattegat, and nearly one hundred pups were born prematurely near the Danish island of Anholt. In May, this was followed by the discovery of a mystery illness affecting all ages of seals off Holland and Germany. Despite this, it was not until well into the summer, with the rapid spread of the disease, that the media began to report on the terrible decimation of the Dutch colonies. Even when it hit East Anglia late in August, it seemed possible that the virus would not travel to Scotland. It puzzled scientists when reports came in from Orkney that the virus was causing deaths there, too.

Holland already had a government-backed seal sanctuary, known as the Seal Orphanage, in Pieterburen. It was Albert Osterhaus, a Dutch virologist from the National Institute of Public Health and Environmental Hygiene in Bilthoven, who identified the virus. He found that it was a species of morbillivirus closely related to canine distemper and measles. As this epidemic represented its first instance among seals, it was given the name phocine distemper virus, after the Latin name for seals. Its effects included suppression of the immune system, thus making way for other diseases to invade.

Many scientists believe that the role of pollution was a major factor in the severity and extent of the plague. Toxins, such as polychlorinated biphenyls (PCBs), heavy metals, DDT and halogenated hydrocarbons (HHCs), contaminate fish stocks and therefore accumulate in the blubber of the seals which eat them. In Germany, scientists discovered over 1,000 different contaminants present in the body of one dead seal pup. During the epidemic, the carcasses of seal victims had to be disposed of in a special way, having been labelled 'hazardous waste'. Experiments have shown that female common seals fed with fish contaminated with PCBs become sterile and have reduced levels of vitamin A, so important in resisting disease. It is also known that pollutants can suppress the production of T-cells which would have been essential in fighting the virus. In the less polluted waters off northern Norway and Scotland proportionally far fewer seals died — another indication that man-made pollution was responsible for the spread of this terrible plague.

In September, George rang Scottish Greenpeace representative Paul Vodden, asking how we could help the seals in the face of this disaster. He offered his support for a seal group in Islay and Jura, and suggested we should contact John Robins, the Organising Secretary of Animal Concern, in Glasgow. At the time, I had written a song for the Islay children's choirs to sing as part of the long-running campaign to have a swimming-pool built on Islay. The children had learnt the song, with the chorus in Gaelic, and in mid-September I conducted them on the steps of Strathclyde Regional Council buildings in Glasgow. Islay now has a wonderful indoor swimming-pool, built in a converted whisky warehouse. It is a credit to this small island community which worked so hard to raise funds over many years; their determination eventually gaining support from the local authorities as well as generous individuals. On my return from Glasgow, I telephoned John Robins and told him how much the seals meant to me and how I wanted to help them. I explained that I sang to them and played the violin on the shores of Kildalton while they swam about and listened. On the one hand, it didn't seem a big thing to me for I was so used to it, on the other, I would rather have kept it to myself because I knew that many people would not understand. I felt John was trying to take me seriously, if only for the fact that he had actually seen me on Scottish Television news conducting the children's choir. Anyway, it must have made him curious, because he decided to take a trip to Islay to see for himself.

I was a little bit apprehensive when we went to meet Mr Robins at the airstrip. I had read about some of Animal Concern's campaigns in the newspapers and John Robins had always been very outspoken over the persecution of seals. As the people disembarked from the plane, I was expecting to meet a slim vegan with dreadlocks hanging out of a balaclava mask. When he introduced himself I was rather surprised. In his somewhat bulging cord jacket and cord jeans he looked a bit like a cross between an ageing young farmer and a refugee from the early Open University broadcasts. On the car journey back to Kildalton, I was pleased to realise he had a genuine concern for seals and wildlife, and a total abhorrence of those who persecuted these creatures.

He had arrived on the afternoon plane and George and my stepson, Peter, came out to the Seal Islands in the small Orkney longliner with us straightaway. The weather was fairly calm but deteriorating with a strengthening wind. Once again I was apprehensive, as I was not in the habit of playing to the seals with other people around, and I knew that if the sea and weather conditions were wrong there would not be any seals at all. I need not have worried, for it was as though they knew the occasion was special, and they turned out in force that evening.

John Robins stayed well back as I played from the rock on the shore of Eilean Bhride and they gathered before me, until a happy throng of over 25 seals were bobbing and swimming up and down in the water, straining their necks upwards as if to get a better view. I enjoyed seeing the look of astonishment on John's face, little realising that this was the beginning of a whole new chapter in my life with the seals.

I had been working on my song 'Today the Seals' in the studio, putting the music together track on track. When I played it to John he was very enthusiastic and urged me to think in terms of releasing it as a single to help raise awareness and funds for the seal sanctuaries in Scotland. I was keen to help but still felt reticent about publicising the fact that I played to the seals. After thinking it over, however, I realised that although it had been up till then a very private and secret relationship, it was important now to do all I could for their welfare. Even if it meant facing ridicule and lack of understanding from some quarters, it did not matter as long as I was getting a message across to those who would be touched by it, and care. At this time the disease had not yet reached Islay, but other parts of Scotland were being affected. I gave John the go-ahead and agreed to work with Animal Concern and make my music available to help the seals.

The first newspaper to take up the story was the *Daily Record*, which dispatched reporter John Nairn and photographer Ken Ferguson from Glasgow. The weather was against us, however, and we spent the best part of two days in the Dower House Hotel looking gloomily out of the bar window on a grey, wet scene and stormy sea. We managed to make some use of this delay and persuaded Ken to take some shots of the children who were campaigning for the swimming-pool. A whole class from the local school posed in their swimming gear in the freezing cold to show just how badly this island community needed proper facilities to teach youngsters how to swim. Ken took some pictures which the *Daily Record* published alongside a story about the swimming-pool project.

When there appeared to be a slight lull in the fierce wind and torrential rain, we braved the rough sea and ventured out in the only small boat available to us at the time. Although it was rocking up and down in the swell, we were safe in the relative shelter of the bay but, due to the conditions, there was not a seal in sight. At one point we all suddenly saw a small, dark head in the water nearby – it turned out to be an otter. When it became clear that there was no chance of getting a picture of me playing to seals, Ken decided that he would find a suitable scenic spot and take a shot without them.

We all went down behind the castle and prepared to take the pictures in the bay between the hut with carved names and the dùn. It was pouring with rain and we huddled in the hut, nipping out to take the shots at the

A stunning silhouette shot was taken as I played by the lagoon at sunset
(*The Telegraph*, plc, London, 1988)

slightest break in the weather. Peter came round in the boat with flasks of
hot coffee from the Dower House. Because of the photograph, I could not
wear suitable clothes to suit the conditions, and I was cold and wet. My
hair kept blowing over my face and I should have been miserable but for
the knowledge that I was doing something to help the seals. When it was
published we were very pleased with the way it turned out. It was a lovely
picture of me sitting on a rock playing the violin, with the sea and Fairy

Hill in the background, and John Nairn's article highlighted the fact that a record would soon be released to raise money for the seals.

I recorded 'Islay Mist' as a violin solo for the B-side of the cassette single. Greenpeace allowed us to use one of their photographs for the cover – a heart-rending picture of a beautiful common seal pup, looking straight at the camera with large, soulful eyes, and a dead adult lying behind him. On 7 November *Today the Seals* was launched along with a press release from Animal Concern about the Save Scotland's Seals Fund which was established to ensure that all money raised would be used for the seals and not other campaigns. It was publicised on Scottish news broadcasts and on the Jimmy Mack Radio Show.

The next day Robert Reid came over to Islay for the *Daily Telegraph* and, once again, there was a problem with the weather. We could not go out in the boat so, instead, we went down to the peninsula where the only seal to be seen was one which was bottling in the waters of the lagoon. Nevertheless, a stunning silhouette shot was taken as I played the violin by the lagoon at sunset, which appeared in the newspaper a few days later. By now people were buying the cassette and I was receiving encouraging letters. After the *Daily Telegraph* article came out, I was amazed when a man wrote to me all the way from Israel in support. *Newsbeat* gave me a report on BBC Radio One and there was an article on the front page of the *Oban Times*. I gave my first interviews on radio over the telephone.

By this time the sanctuaries in Orkney and Skye were very busy and dead seals were being found up and down the west coast of Scotland. The Islay and Jura Seal Action Group now had all the available information at hand in the event of being faced with a sick seal. We were liaising with the RSPCA and Greenpeace at the newly built rescue centre at Docking in Norfolk. Here seals suffering from the virus were being successfully treated with antibiotics and other medicines, and a new vaccine, which had been developed in Holland, was showing hopeful results. We had been in contact with Ross Flett and Maureen Bain on Orkney, and Grace and Paul Yoxon on Skye. Although hundreds of seals were dying, there was a glimmer of hope as some began to recover from the effects of the virus, becoming immune to the disease and so building up a resistant stock for the future.

The *Daily Mail* had been running a marvellous campaign, with huge support from concerned readers, and we were offered funds for a rescue boat. On the mainland, Don and Dot took delivery of a 17-foot Osprey Sparrowhawk – a hard-bottomed inflatable that would enable us to go out safely in the treacherous waters off south-east Islay. The boat was launched from the Dower House Bay in the presence of photographer Jim Hutchinson from the *Daily Mail*; the ceremony being performed with a bottle of local malt whisky. Everyone in the group was delighted and relieved that we now had a suitable boat to work with, and we were extremely grateful to all the generous *Daily Mail* readers. A picture of the boat appeared in a four-page pull-out in the newspaper, featuring all the various seal groups that had been helped by the *Daily Mail* appeal.

After reading the article in the *Daily Telegraph*, a reporter from TVAM telephoned us seeking an interview. We spoke with Lorraine Kelly who, having been brought up in Scotland herself, was very keen to cover the story of a woman who played to the seals on Islay and who had brought out a record for them. With a crew of cameraman and sound recordist, she travelled by plane to Islay in the middle of November. John Robins joined them and flew over from Glasgow to be interviewed at Seal Bay.

I had told Lorraine of how I would go with Jason round the shore and she liked the idea of filming me riding to visit the seals. The crew were driven as near as possible to the chosen site and then had to carry their heavy equipment across extremely wet and muddy terrain. As the crew unloaded their very expensive gear from the car, they realised that they did

We rustled up some dustbin liners and rubber bands to keep their feet dry
(Animal Concern)

not have one pair of wellingtons between them. So we rustled up some dustbin liners and rubber bands to keep their feet dry as they filmed me riding Jason across the sandy bay and round to the rocks where an audience awaited. I then sat by the sea and played the violin to the seals – although, with other people being there, the atmosphere was different from my usual relaxed and natural concerts. A few seals came into the line of the camera as I played, but it was not easy to capture the whole feel of the scene in a few moments and Lorraine decided that more shots of seals should be taken separately.

We gathered at the Dower House where I was interviewed on the lawn by the sea. I spoke of my love for the seals, my concern for them and how they needed help. Don then took Lorraine and the crew out in the new boat – the costly camera protected with yet more dustbin liners – and they filmed groups of seals in the water and lying on rocks among the islets and skerries beyond the Fairy Hill. Although the sea was still a little choppy from the storms, it was a bright day and certainly better than on the previous occasions with the press. Lorraine thoroughly enjoyed going out in the boat and was thrilled with the seals. She was obviously interested in their welfare and wanted to help us in our efforts to protect them.

On the morning that I was to be on TVAM, we woke up at six – not knowing exactly when Lorraine's slot on the seals would be on. Just as we were wondering if it had been pushed out by some other feature, the presenter introduced a report from Islay. As I was shown riding Jason in Seal Bay and playing 'Islay Mist' by the water, Lorraine's commentary talked about the seals to whom I played music and of how they were threatened by the virus. She spoke of the record *Today the Seals* and how I hoped that it would raise funds for the Scottish sanctuaries; and all the time my music played as seals were shown in the water and hauled-out resting. John Robins stressed the lack of care available for the large population of seals in Scotland. When the piece was finished, the presenter and a guest in the TVAM studio that morning commented on how appropriate the words of the song were – 'Today the seals, tomorrow who'll suffer while mankind plays the fool?'

The next day, the *Independent*'s columnist, the Weasel, had an amusing cartoon drawing of me playing the violin to a seal, with a paragraph to explain what I was doing for the seals. Far from being ridiculed, I was capturing people's imaginations. A couple of days later, as part of their continuing campaign, the *Daily Mail* ran an article on my record with a photograph that Jim Hutchinson had taken of me when on Kildalton for the boat launch.

Healthy youngsters rode on their mothers' backs

Rocks thick once more with dozing seals (Chris Davies)

87

During the next week, I was up until the early hours sending letters and cassettes of the song to Members of Parliament, including Prime Minister Thatcher. The letter asked Parliament to take action to support the small local sanctuaries on Orkney, Skye and Islay, and to put a complete stop to all shooting of seals in Scotland, giving them the same rights as in England and Wales. A temporary ban was put in place during the epidemic, but it was not enough. A lot of effort goes into saving each individual seal only to have it released back into the wild where it could be shot. Animal Concern had been campaigning on this issue for some time. I ended the letter by saying that I wanted my daughter, Hannah, to have seals to play to when she grew up.

I was busy working on the campaign, and the seals were being monitored off the coasts of Islay and Jura. We had come off extremely lightly, as only 28 seals were found dead here – although it was probable that more than that had died but had not been washed up – whereas, just over the water on Kintyre and in the Clyde, many hundreds were lost. Despite our preparations and efforts, the seals were all dead before we reached them, but we had gained valuable knowledge which would stand us in good stead for the future. Some scientists thought that the virus would die out and not return; others warned that the epidemic could be repeated in years to come.

In February 1989 there was a meeting in Inverness of Scottish Seal Rescue groups which George, Peter and John Robins attended. In particular, George was pleased to meet Ross Flett from Orkney with whom he had spoken many times on the telephone and, through their communication, useful information was exchanged between the groups. In a bar that night, George, Peter and John met some fish-farmers who had been attending a fish-farm conference in Inverness. Unaware of who John Robins was, they told of the killing of seals and other wildlife that went on around the Scottish coast.

In the spring, we went out in the Seal Patrol Boat as the pupping season began. We watched as healthy youngsters rode on their mothers' backs and played peacefully in the shallows in the warm sun. The rocks were thick once more with dozing seals and it seemed that the nightmare was over. But I had learnt a lot about the continued persecution of seals in Scottish waters and, because of my love for them, I could not let the matter rest.

SIX

People of the Sea

'What about a grand piano in the sea?' George suddenly asked me with a big grin on his face. The very idea horrified me. I had been brought up to respect and care for musical instruments, and anyway, where would we get one on a Hebridean island? The only grand pianos I knew of locally were given pride of place in people's homes.

George and I had been discussing ways of raising public awareness in the campaign against the slaughter of seals in Scotland. I had already made up my mind that his crazy proposition was impossible, but I should have known better. Within a week I had a telephone call from Edmund Lawless, a fellow musician living on the estate. A violinist and pianist, he lived in a cottage between Quartz Lodge and the Fairy Hill. He was moving to another part of the island and wondered if I would like to buy his grand piano. I knew he had gone to great expense to have it French-polished after his house-warming party got a little out of hand, and was afraid that I also knew what its destiny would be if I did buy it.

In 1978 there was a public outcry at the proposed seal cull in Orkney which put a stop to any official mass slaughter. At the time of the virus in 1988 public sympathy was firmly behind the welfare of the seals, with hundreds of thousands of pounds raised throughout Britain for the care of sick individuals. For many months the *Daily Mail* ran a very successful campaign, full of touching seal stories and articles about scientists who were trying to find out the cause of the virus and how to treat it. One would have thought that the seal was a much loved animal, safe in British waters.

But with the rapid growth of the fish-farming industry on the West Coast of Scotland since the mid-1970s, a new insidious threat to seals and other wildlife emerged. By 1991 there were 327 marine fish-farming sites between the coast of Argyll and Shetland. Millions of salmon are now packed into small floating cages and forced to circle away their lives, whereas in the wild they would normally migrate thousands of miles through the oceans. Sited

in once beautiful sea lochs, they pollute the area with faecal and food material, up to 25 per cent of which falls to the seabed. Chemicals and additives are used to ward off disease, kill parasites and to give the grey flesh of farmed fish a more attractive pink colour. Moreover, there is a threat to wild salmon through disease and genetic degradation by escaped and released domestic stock.

Anything that eats salmon is the enemy of the fish-farmer and salmon-netsman and is, therefore, liable to persecution and destruction. As long as they use the correct calibre of bullet, salmon farmers are allowed to shoot seals. The Conservation of Seals Act 1970 prohibits the shooting of seals in Scotland during their respective breeding seasons – for the grey, this is between 1 September and 31 December; and for the common, between 1 June and 31 August. But if you are a fish-farmer or a netsman, you may still obtain a licence from the Secretary of State to kill seals at these times. The wording of the Act states that the killing of wildlife is allowed to 'prevent serious damage to livestock', but as this is not defined, it is left wide open to individual interpretation. While there remain a few fish-farmers who have a policy against shooting seals and other wildlife, the majority regularly kill as a method of predator control. Not only do they shoot at seals seen near cages, they have been known to go out in boats to

OPPOSITE:
John Robins of Animal Concern holding a common seal pup shot on the river Helmsdale (Animal Concern)

Ross Flett of Orkney Seal Rescue carrying a decapitated seal found near a salmon farm (Orkney Photographic)

91

shoot seals at haul-out sites – in one incident at a haul-out which was two miles from the fish-farm. In another incident five men went out in a boat with at least three guns and shot indiscriminately at the wildlife in the loch. There is a known case where a seal was wounded with a shotgun. An entire heronry was destroyed with a shotgun and, in one day, seven seals were killed near a fish-farm in Orkney. At another fish-farm six otters were killed during the course of a year.

There are no accurate records of the precise number of seals and otters that are shot each year, or for the thousands of shags, cormorants, herons and gulls. The Marine Conservation Society has reported that some operators deliberately set their nets loosely so as to tangle and drown seals. Bounties are known to have been paid to men to shoot seals, and one salmon-netsman admitted, quite openly, to shooting 91 seals in a year, calling them 'just vermin of the sea'.

Is all this in keeping with the image of Scotland we see presented in tourist board advertisements? A seal which had been shot twice in the head but was still alive was seen by boat-loads of tourists cruising near Oban. Shooting seals often means a lingering, agonising death. Are people aware of the price wildlife has to pay for the salmon they buy? If not, I am determined to keep reminding them. And the MPs in Westminster, who so readily condemn the seal culls in other countries such as Norway and Canada, should take a long, hard look at what is happening on their doorstep. In parts of Canada and the USA, it is illegal for fish-farmers to kill predator species. Animal Concern estimates that around 5,000 seals are being shot each year in Scotland.

There are many other ways of protecting stock, including different types of netting, sonar devices and even models of killer whales which are suspended in the water. But the costs are considered prohibitive. If the Crown Estate Commissioners insisted on these requirements in the seabed leases they issue, or if a law was passed and properly enforced to protect seals and other wildlife, the operators would be forced to use the many humane deterrents which are available. Effective, compassionate stock protection would, inevitably, benefit the fish-farmer whose stock would be continually secure. At present, many farmers react to problems *after* they occur and damage has been sustained.

Every major sea loch in the West Coast of Scotland – home to the seals for thousands of years – now has at least one fish-farm. The waters of the people of the sea have been encroached by man, and the seals are being shot for living there.

Thinking over the terrible things that were happening to seals in Scottish waters, I knew that we had to go ahead with George's plan of using the grand piano. I felt sure that Edmund would forgive us when he realised that his precious instrument was being sacrificed to a good cause. We would take as much care as possible to bring it back from its little sojourn in the sea without damage.

We had to choose a site that could be accessed by vehicle, but which, at the same time, had plenty of sea surrounding it. The tidal island was decided on as the best place as the ground would be reasonably hard in the summer. Help was needed from friends and family. At low tide, Donald MacKinnon, the brewer from Lagavulin distillery, his son, Nicol, and Peter crazily manoeuvred the piano, which was lashed to a four-wheel-drive vehicle, over the rough ground of the peninsula and across to a little islet off the main Imersay Island. Bales of hay were used as padding and protective coverings kept the piano dry overnight. TVAM were due to arrive in response to a press release sent out by Animal Concern. This concert, highlighting the persecution of the seals, was to be a serious affair. My brother, William, flew in from Germany to perform and, because the piano would be out of tune after the removal, our piano-tuning friend from the Sevenoaks days came with him. Francis Rwama from Uganda was a much-loved musician whom we had known from his days at Dorton House School for the Blind, near Seal. I had often accompanied him on my violin as he played jazz on the piano. The evening before the television crew was due to arrive, William and I practised some pieces together and I showed him how to play the piano part of 'Today the Seals' for me to sing to.

TVAM arrived on the morning plane with Louise Bevan as reporter. The schedule was planned so that Don, George and I immediately took the television crew out to the Seal Islands in the rescue boat to take some shots of the seal sanctuary. Meanwhile, Francis was taken down to the peninsula island to tune the piano as friends and family were gathering for the concert. There was a wind blowing, but the weather was mainly dry.

Don dropped us off on Eilean Bhride and, after a bit of a struggle over the rocks and uneven ground, they found a good spot and began to film the seals. After a few minutes, the cameraman began to look unhappy and started tinkering with the equipment. Inexplicably, something had ceased to function and he was unable to fix it. After a while, we all got back in the boat and returned to the Dower House where further efforts to mend the camera also failed. It was thought that possibly the damp had affected the electrics and they tried to dry them out with a hair dryer. This too was to no avail. A few phone calls and several cups of coffee later, it was arranged that a new set of equipment would come in on the afternoon plane. Meanwhile, everyone else was waiting patiently on Imersay Island in the cold wind.

At last, the new equipment arrived and I walked across the sand to the islet dressed in my long evening kilt – done up with safety pins and with a suitably loose top as I was nearly seven months pregnant. One microphone was set up in the piano and another pinned to my clothes. The wind was blowing my brother's music off the stand so we had to stick it down with brown tape. TVAM particularly wanted 'Today the Seals' for their report so I sat on a rock and sang while William accompanied me, smartly dressed in his black suit and bow tie; coat-tails neatly flowing over the back of the piano stool. I was then interviewed by Louise, saying how seals should be loved and respected and not treated like vermin.

By the time all the recording had been completed the sky was growing

My brother flew in from Germany to perform and our piano-tuning friend, Francis Rwama from Uganda came with him (Animal Concern)

dark and the tide was coming in, cutting the island off from the peninsula. Most of the party managed to wade across, myself included with kilt held up above the water. But Louise Bevan and the television crew, fearing more damage would occur to the equipment, waited on the island for a boat with George, John Robins, William and Francis. As blustery showers came in from the north-west, they tried to shelter under the grand piano. As darkness fell, William played 'Blue Moon' and seals stared in disbelief at the bizarre ensemble marooned in the gathering gloom. The seal boat, piloted by Don, finally came and rescued them. He had been a little delayed as he had been busy serving dinner at his guest house in Lagavulin. Of course, if he had been called out to rescue a seal, the guests would probably have had to wait!

That day had been reserved for TVAM and the next day the press had

William picked up *The Scotsman* and gave a startled chortle of delight. There he was on the front page in coat-tails and rolled-up trousers by the grand piano with the headline caption 'Pythonesque Ploy to Highlight Plight of Seals' (*The Scotsman*)

been invited. George and John Robins went to the airport to see the television crew off on the morning plane, dropping me off in Port Ellen to pick up some shopping. They watched as people disembarked from the Glasgow plane and were disappointed not to see anyone from the newspapers. It seemed that nobody had come to cover the story. Just then, they heard another drone from the skies and a light plane landed on the tarmac. The press piled out from the private aircraft which had been piloted by freelance journalist Tom Kidd.

I was surprised to be picked up from outside the shop in Port Ellen, with my leeks and tomatoes, by a white minibus full of newspapermen. However, it was nice to see someone I already knew as I recognised the photographer who had taken the silhouette photograph for the *Daily Telegraph*. We took them to the farmhouse where they were fortified with the local brew before trekking down to the tidal island.

When the covers were taken off the piano, George noticed a crab scuttle out from the strings. As Francis had a quick play on the piano to check the tuning, the reporters stood about looking a bit nervous – it was as though they thought Jeremy Beadle was about to pop out from behind a rock at any moment! William and I played a wide variety of music as the photographers clicked away from all different angles. They loved William stepping out to the semi-submerged piano with the trouser legs of his tailed suit rolled up, then sitting on a piano stool that was threatening to float away. Little curly-haired Hannah was greatly excited by the whole event, and my younger stepdaughter, Sarah, was looking after her while I played. Unfortunately, Hannah tried to copy mummy and, when I put my violin down for a break, she broke my best bow in two. The piano was covered up once more and everyone went back to the farmhouse for further sustenance.

The next day, William and George went to Port Ellen to get the papers. William picked up a *Scotsman* and gave a startled chortle of delight. There he was on the front page, standing by the grand piano in coat-tails and rolled-up trousers, with the headline caption 'Pythonesque Ploy to Highlight Plight of Seals'. We also generated a lot of other press coverage, including the front page of the *Daily Express*. This nationwide publicity was a great help to Animal Concern's campaign to stop the slaughter of seals.

A BBC radio programme on the conflict between fishermen and seals was being made by Jessica Holm for *Wildlife on Four*. She arrived on Islay to record me playing the violin to the seals and I was to be interviewed about the unofficial cull going on in Scottish waters.

It was beautiful summer weather and we decided to take Jessica out in the boat to the Seal Islands where she could see the seals with their pups.

As we approached Eilean Bhride we slowed the boat and I got out my violin and sat at the prow. Jessica prepared her recording equipment and, as I played, we rowed forward into the wide channel between the two main parts of the island. Jessica became very excited in her descriptive running commentary as seals porpoised and frolicked all around us. It was a lovely moment and I was pleased that Jessica was there to share it with us.

In the programme she told of the thousands of seals being slaughtered in Scotland in an unofficial cull. She had recorded me playing 'Islay Mist' down on the peninsula and this was used in the trailers for the programme. I was glad to have taken part; to have done something to defend the seals in the controversy by bringing out into the open the truth that such a large number of seals was being killed – a fact that some had been trying to conceal.

International freelance photographer Chris Davies from the Network agency was driving to the Royal Academy of Dance in London to take pictures of ballerinas when he heard the programme on the radio. He was so impressed by the scene described by Jessica as I played to the seals that he thought, 'I must go and take photographs of this.' Not knowing how to get in touch with me, he first rang Greenpeace who put him on to Orkney Seal Rescue. At last, he was able to telephone us and told me that he wanted to help the seals with his photography. A couple of days later he was on an aeroplane on his way up to Islay. *Life Magazine* had given him an assignment to take photographs of me playing the violin to seals.

I was impressed by Chris's total professionalism. He was so keen to get the perfect shot in the right light that, as far as he was concerned, the day started at 6 a.m. if necessary, and only finished when his light meter told him so or the sun had set.

When Chris first arrived, George took him up the Fairy Hill to see the view across to the Seal Islands and told him earnestly about the fairy legends. Chris politely said he had an open mind. Then George recounted a recent incident in which the Ministry of Defence commandeered a patch on the hill while conducting a major naval exercise without consulting the Fairy Queen. The most up-to-date surveillance equipment – a microwave transponder – had been installed at the trig point at the top of the Fairy Hill. Fearful of this disturbance to the fairies and the effect it could have on them, George dismantled it and carried it down the hill, locking it in one of the outbuildings. He let the Ministry of Defence know that if they wanted it back, they would have to pay a nominal ransom of a hundred pounds as a donation to the Islay swimming-pool fund. Islay still had no swimming-pool at this time and children had to learn to swim in the sea. When he refused to hand over the transponder, George received a rather formal

He was so keen to get the perfect shot in the right light that, as far as he was concerned, the day started at 6 a.m. if necessary, and only finished when his light meter told him so or the sun had set (Chris Davies)

phone call from a Lieutenant Atkinson from the Admiralty who was not impressed with George's claim that the equipment could have been causing the fairies headaches. There was a destroyer and other Navy ships out in the Sound which were unable to proceed with their exercise. However, in the end, after several hours of negotiation, the Navy saw sense and a cheque was hurriedly produced for one hundred and twenty pounds – the slightly larger amount in recognition of the delay – accompanied by a letter of apology for the disturbance as requested by George.

Sometimes they were terribly jagged and not at all suitable for a pregnant lady to
sit on

George Robertson MP was staying at the Dower House at the time and
found the whole incident very amusing. A few days later, a picture and
headline 'Don't meddle with the Fairy Queen' appeared on the front page
of the *Herald*. It informed how the Ministry of Defence had learned that it
didn't do to ignore the Queen of the Fairies. When it had set some vital
equipment up on her hill without even the courtesy of asking permission,
it was carted away by her 'Ambassador' who then demanded a ransom. On
an inside page was the headline 'Queen of the Fairies Vents her Wrath' and
the full fairy story.

99

A bottle of Islay malt and a tumbler had been placed on top of the hill in deference to the Fairy Queen (*The Herald* and *Evening Times*)

During the next few days I went out to the Seal Islands with Chris and George and spent hours playing to seals while Chris took photographs. Don strapped a full-size armchair to the back of the pilot's seat of the boat for my comfort, as I was heavily pregnant. The islands span several miles and, as we moved from place to place looking for the best locations, I sat and played on many rocks which may well not have been touched by humans for thousands of years. Sometimes they were terribly jagged and not at all suitable for a pregnant lady to perch on, but I tried to accommodate Chris's artistic requirements as far as possible. For the most part, the weather was beautiful, and by spending all day out at the islands, I learned even more about the ways of the seals.

The character of an area could be completely altered between high and low tide. When the sea was well out, pale, exposed rocks lay strewn with slippery stranded weed, and the blue water of summer days rested at their feet. Then, as the tide turned, it crept with stealth to reclaim the land once more, and now only the bare black tops of the rocks could be seen. The sea would become darker and a swell stroked the islands.

There were more suitable rocks to find as seats, but the seals, who were constantly on the move, would be more willing to come to these isolated ones and listen to my playing. It was much harder to take these photographs than was at first thought, for the seals did not like to come between myself and the camera. Nor were they enamoured by the tripod, and the click as Chris pressed the button sent them underwater as though they thought it was a gunshot.

Chris took some photographs in beautiful pink and gold Hebridean sunsets – some of the best pictures he had ever taken. One of these was of me playing, with the Fairy Hill in the background, when a seal popped up to listen. After four days' shooting, he returned to the city lights. He phoned to let us know that the pictures were superb. But a day later, Chris called again very upset. For the first time in his professional career he had lost commissioned photographs. They had disappeared out of his car in his briefcase. Would I mind doing it all over again?

When the photographer from the *Herald* had come to take his pictures a bottle of Islay Malt and a tumbler had been placed on the trig point on the top of the hill in deference to the Fairy Queen. On his forced return, Chris agreed that the first thing he should do was to visit the Fairy Queen and gain her consent for his commission.

And so we climbed to the top of the Fairy Hill and, as Chris snapped away with Hannah and myself playing the violin, a rainbow appeared over the islands as a seal of approval. He took an even wider variety of pictures for his portfolio. Linking in with our campaign work, his photographs have been published all around the world – in this country they have appeared in a full double-page colour feature in *Woman's Own*. Chris is now a firm believer in the magic of the Fairy Hill.

With the success of the 'Today the Seals' cassette single, John Robins had persuaded me to record more of the songs I had composed while playing to the seals to help with Animal Concern's campaign. Over many months, often working in the studio until the early hours of the morning when I would finally be disturbed by the cries of peacocks and cockerels, I produced the *Today the Seals* album.

In my studio, with the walls covered in seal posters, I composed all the parts, recording them track upon track, and finally singing the vocal. (I used the adjoining bathroom to create natural reverb.) Producer Derek Wadsworth came up from London to give me valuable advice on recording technique and

And as Chris snapped away with Hannah and myself playing the violin, a rainbow
appeared over the islands as a seal of approval (Chris Davies)

let me have his sequencer. This enabled me to synchronise the instruments I
played on the synthesiser and sampler. Derek enjoyed his stay at Kildalton,
visiting the seals on the islands with Peter in the Orkney longliner.

One of the tracks on the album, 'Melody in Green', which I had written
not long after arriving at Kildalton, turned out to have a strange connection
with the past. The Seal Islands had been kept by Violet Clifton when the
estate was sold and were still in the Clifton family. When we heard they
were up for sale, we were concerned and thought that the best way to
protect them – as they were so important for the seals – was to buy them.
George visited Melody Clifton in England and learned she had spent her
courting days in Kildalton woods. Her maiden name was Green.

When all the songs had at last been mixed, we prepared for the release
of the cassette and picture CD album. Animal Concern was involved in the
publicity for the campaign to protect seals. A song book with campaign

His portfolio now included a family picture taken in our farmhouse courtyard along with Sandy standing with Misty and the haycart (Chris Davies)

literature was also produced. I appeared on television programmes including *Sky News*, which is beamed around Europe, and the children's programme *Motormouth*. When I talked of the campaign to Radio Foyle, 'People of the Sea' was played in the background. West Sound radio played two songs from the album every week, explaining that the record had been released to highlight the killing of seals in Scotland. And the national press took up the story with features and photographs. The message being given was clear: Scottish farmed salmon means dead seals – please vote against this slaughter with your purse.

I received many wonderful, sympathetic and supportive letters – a large number of them addressed simply to Fiona of the Seals, Kildalton, Islay. Many people wrote directly to me requesting the album. They had no trouble reaching me on the island, and I was amused to hear that local people were dressing up at Hallowe'en as Fiona of the Seals, complete with violin and 'seals' in tow. One letter was from a man called Kurt Rose from Germany who was writing an oratorio on man and the environment which was to be performed in a cathedral. As part of his illustration of man interacting with nature, he described a scene in which I played in the tranquillity of an unspoilt, wild place to the seals.

While I was still working on the album, John Robins had come over to Islay with George Hume and a BBC crew to film a feature for the Scottish documentary programme, *Hume at Large*. I was filmed with Hannah as I played to the seals and in my studio, and George was shown walking amongst the thick bracken on the Fairy Hill to the background music of Tchaikovsky's 'Sugar Plum Fairy'.

At this time I was looking after the terribly sick pup which had been found in the Dower House Bay. Tarzan appeared briefly in the film, with Hannah and Fifi watching me nurse him. He was given all the love and care possible along with expert veterinary advice from Janet Berry, but sadly he did not live. As he fell into his last deep sleep late one night, I spoke on the phone to Ross Flett of Orkney Seal Rescue. He and Maureen Bain had given all the support they could, sharing their practical knowledge. Having experienced the joys and sorrows of looking after seals, Ross's understanding helped me through the pain of losing Tarzan. I was to face such grief again and again, but success would also come with the elation of returning happy, healthy animals to their ocean world. One day, I hope, this emotion will no longer have to be tempered with the fear and uncertainty which is caused by the knowledge that seals travel far and many fall foul of man's evil intentions.

After Tarzan died I sat at the grand piano and wrote 'People of the Sea'.

PEOPLE OF THE SEA

Ancient stories tell,
seal maidens from the sea,
searching for a love,
longing to be free.
Some would swim ashore,
there to walk with men,
returning to the sea
to live as seals again.

Sad tales are told
of seals who sing,
lamenting the pain that man can bring.

People of the sea,
when I heard your cry to me,
and I knew that many'd never understand.
Well I said I'd fight for you,
with the ones whose hearts are true,
with those who love the people of the sea.

Legends from the past
tell of drowning men,
safely brought to shore –
selkies rescued them.
Stories of today,
seals in fear of man.
Where now is your home?
When will they understand?

God gave you the ocean,
these are yours –
the waters of the world.

People of the sea,
your dark eyes plead with me
though I'll never know the way you really feel.
But if any harm comes to you, I just feel I'm hurting too,
because I love the people of the sea.

People of the sea,
when I heard your cry to me,
and I knew that many'd never understand.
Well I said I'd fight for you,
with the ones whose hearts are true,
with those who love the people of the sea.
I'm not ashamed to love you, people of the sea.
People of the sea.

SEVEN

Aqua

It was a beautiful warm day in June, and with the kitchen door wide open as I made up sandwiches, we heard Don holler from the bay below. After gathering up food, children, violins and binoculars, we made our way down the woodland path to the pier where Don was waiting with the seal-rescue boat. We headed out across the clear blue water, passing groups of seals lazing on the outlying skerries and, as we neared the seal sanctuary, Hannah and Fifi pointed excitedly at mothers with pups riding on their backs.

Don manoeuvred the boat against some low rocks by the side of the channel and George helped the children out on to Eilean Bhride. The balmy scent of wild flowers mingled with a salty sea breeze that carried the busy,

Even a small wave would have washed me off that particular rock

OPPOSITE:
Aqua followed at my heels like a well-trained dog

muffled chatter of cliff-loving gulls from distant rocky isles. But here on Eilean Bhride all was quiet – save for the ever-present breath of the living sea and the occasional piercing cry of a passing oyster-catcher.

Don carefully pushed the boat away from the rocks with an oar and made his way out of the channel to check on the health of the widespread seal colony. We chose a grassy spot amongst the boulders and rock pools for our picnic. Hannah and Fifi were soon busy finding flowers and feathers, pebbles and crab claws to play with, the empty shells dry and brittle from the hot summer sun. I wandered to the end of the island and played my violin to the seals across the lagoon. There were several mothers watching their pups playing happily in the shallows. Others lay contentedly on the rocks by the sea edge. As I returned to the picnic, Peter turned up in the Orkney longliner, and he and George left to scan the string of small isles just north of Eilean Bhride and over to Outram.

As I sat watching the children play with their new-found toys, I thought of all the different places I had played to the seals from up and down these miles of skerries and isles. What a perfect habitat it was; giving them shelter in the pupping season and peace and safety in the summer months as the little ones grew and learned how to fend for themselves. How sorry I felt for those colonies that are harassed or even wiped out by fishing interests. Here, the seals were secure in their own kingdom, and they were unafraid, having come to know us just as harmless curiosities occasionally passing through.

I laughed to myself as I remembered how George and Chris had left me to play on a tiny rock by the open ocean as they attempted to take a photograph from a different perspective. As they rowed a distance away in the boat, George told me with a grin about the passing killer whales who might mistake me for a tasty seal and of the submarines that ply the Sound creating huge tidal waves. Even a small wave would have washed me off that particular rock. There are whales in the sea here but I knew that the possibility of an orca just happening to be passing at that moment was very remote. Sperm and minke whales have been washed up dead on the Atlantic side of Islay, as has a striped dolphin which we helped dispatch to the strandings research unit in Inverness. Peter once saw a whale off Churn Island and he, Don and Dot have occasionally seen porpoises when out fishing. On one occasion, Peter had eight of these beautiful animals playing around his boat. But, sadly, it seems that they are less common now than 15 years ago.

Hannah got out her tiny violin and scratched away happily. In the distance, a few seals looked up from their nap for a moment, then settled down again to sleep. They probably just thought that one of the seagulls was feeling ill. Fifi looked earnestly through a pair of binoculars that she had made with a couple of toilet rolls, coloured paper and string. I was just thinking that George was taking a lot longer than usual, when he returned with Peter and we made our way back home.

When we were in the house, George went away and came back telling me that there was something wrong with the plumbing in the bathroom of the spare bedroom downstairs. I thought it was a little odd but I knew the toilet cistern had a habit of overflowing now and then. So, being 'Mrs Fixit' in the household, I went to investigate. As soon as I entered the room I knew what was up. The salty, seaweedy aroma hit me as I entered the bathroom. A gorgeous little seal pup lay in the bath looking at me with bewildered dark eyes.

I rushed back to George and, after telling him off for teasing me, asked what seemed to be wrong and where he and Peter had found her. Peter had seen her earlier on in the day and was concerned as she was on her

Hannah got out her tiny violin and scratched away happily

own. As they approached the 'seal hospital', they saw that she was still lying alone, looking lost amongst a mass of green seaweed by the sea edge. Believing that she may well have been abandoned by her mother, they went in close with the boat. She did not appear well at all. Pete stepped out on to the seaweed but she made no effort to get into the water. He picked her up and put her in the boat where she lay very quietly, unusually lethargic on the trip back to the pier. We have learned from experience in our sanctuary that the common seal mothers never leave very young pups unattended for long, and they are always either lying on the rocks or kelp or swimming in the sea together.

I put the kettle on to sterilise some water and phoned the vet for further supplies of antibiotics and oral rehydration powders. The first thing we had to do was to give her some of this special combination of energy-giving glucose and body salts mixed in water. I put the stomach tube and syringes into a plastic tub with sterilising tablets, just as one would do for a baby's bottles. When all was ready, I crushed up an aquatic animal vitamin tablet and an antibiotic pill and added it to the liquid mixture.

Hannah and Fifi were excited at the new arrival, but they knew that I went to feed seals by myself, especially in the first few days when they were settling in. I took the tube and syringes in a bowl to the bedroom and prepared a space on the floor with some old towels. Some sanctuaries use a tube with a funnel to pour the liquid down, but since I usually have to

He picked her up and took her to the boat

OPPOSITE:
They saw her still lying alone, looking lost amongst a mass of green seaweed

work on my own, I find the syringes best. I went to the bath and, putting a towel around her body, picked up the seal pup and put her on the floor. She was very young, only days old, and the stump of her umbilical cord was still reddish pink from the birth. It would need to be sprayed with a special antibiotic, for if infection set in there, it could be fatal.

She did not try to bite me but I had to hold her still, keeping her fore-flippers under control, so I sat firmly astride her back and held her head steady. So as not to stress her, I tried to do everything quietly and calmly and get the feeding over as quickly as possible. With my gloved left hand, I gently opened her mouth from the side and slipped the tube down into her stomach with the other hand. When Janet Berry had first shown me how to do this with Tarzan I had found it difficult. It is easy to feel you are hurting the seal in some way, but after learning that the length of the tube ensures that it reaches the stomach and does not go into the lungs, and realising that it is something the seals very quickly get used to, I became quite comfortable about doing it. Once the tube is in place, with only a short piece left sticking out between the front teeth, the syringes can be slotted into the tube one after the other and the liquid delivered. The seal's head has to be kept steadily under control. If not, she would be able to shake her head from side to side, flinging the tube out again.

After gently removing the tube, I placed her back in the bath on a couple of old towels. Her breathing was wheezy and so far she had made no noise at all. This was unusual as young pups normally make a sound which we have come to call their 'woo'. I put the fleshy part of my thumb by her mouth but she was not interested in sucking. She seemed to have breathing difficulties – possibly caused by infection – coupled with a lack of desire to suck and an apparent inability to call, which could have caused her to lose contact with her mother.

Having made her comfortable, I returned to the kitchen and began making a record of her feeding programme. She would have to have four or five feeds a day. As she was so small, I decided on five, one of which would double as a visit during the night to check on her as well. I had mixed feelings about the arrival of another seal in my care. I knew that if she had been left out at the islands on her own, she would certainly have died from starvation and, as she grew weaker, the gulls may well have attacked her and caused further suffering. However, the methods used for bringing up young seal pups have only been developed in recent years. I would do my very best to rear her and prepare her for release back into the wild, but I knew that there would be many hurdles to overcome, and I also had experienced the grief of failure.

Another abandoned premature pup still had thick lanugo fur covering his entire body

There had been Tarzan, who was so small and weak. It appeared that it was gastroenteritis that finally killed him, but he was probably premature. I had looked after a female common seal of about six months who had been found on the Atlantic shore on the other side of Islay after a storm. She had very bad internal injuries but the vet could not determine their extent, and I nursed her for several days before she died. It was only then that the Dutch vet, Paul van der Heiden, could confirm the worst scenario at post mortem. Another abandoned premature pup still had thick lanugo fur covering his entire body. Only his beautiful little face was bare. Normally, this fur is shed in the womb at birth and is generally thought to be white, but this tiny pup's unusual covering was grey. The fur gradually fell away to reveal his true coat. He was so happy and contented in his short life. He cleaved to me as though I was his own mother and I did all I could for him. On the whole, he seemed to be progressing well, but I was always concerned that his temperature swung up and down, and it was a constant

Ellen was released in Seal Bay to a serenade of 'Islay Mist'

worry trying to keep him on an even keel. To warm a seal, a heat source
of some kind has to be provided, and to cool one, the flippers are wetted.
One day, despite recent veterinary attention as he had seemed unwell for
several days, his temperature soared; he became gradually lethargic and
died. We had had him for several weeks. The whole family was devastated,
and I couldn't help crying for days.

The first success came with a six-month-old common seal which had
been looked after by Don and Dot in their guest house at Lagavulin. In
December, I had been alerted to the frequent presence of a seal in the bay
at Port Ellen. I went down to the pier, where the fishing-boats tie up and
the ferry docks, with George and Hannah. We saw the seal swimming close
inshore, and observed that she swam with her back above the water and
only took short dives. I knew this meant it was likely that she had a problem
with lungworm, but there was nothing we could do while she was wild and

free. On New Year's Day, Don and Dot responded to a telephone call and found her very ill on the shore near the pier, 'being sick at both ends'. Paula, Janet Berry's assistant at the time, attended to the seal which had been named Ellen. She was successfully treated for worm infestation and, with Don and Dot's care, she recovered. After a time, she was moved from one of their guest-house baths to a large tank that Don acquired for her, where she put on weight and became healthy and ready for the wild again. She was released in Seal Bay to a serenade of 'Islay Mist', and the photographs that Chris Davies took of Ellen were used on the *Today the Seals* cassette album cover and picture CD.

We named the new seal pup Aqua. For the first week I took her temperature every day and, apart from a slight rise at the beginning, I was relieved to find it remained steadily normal. This emphasised to me that all along there had been something badly wrong with the premature pup that had died. Aqua was unable to call for the first week, but her voice gradually came back, hoarsely at first, until in the second week she sounded just like any healthy pup. I would hear her calling from the bath and felt sorry for her being lonely. She could hear me as I hung up washing outside her window and the 'wooing' would start up.

The feeding had gone well. After the first day of liquid-only feed, small amounts of liquidised herring were added to the mixture and this was gradually increased each day. Then, after a few days, small slices of fish were force-fed over her tongue. These amounts were gradually increased while the liquidised fish was decreased. This regime is necessary in order to get the pup's digestive system used to the unnatural diet. In the wild she would be getting her mother's milk only, which is particularly rich in fat and unlike any other mammal's milk.

After the first day, Aqua became more lively and would fight and struggle at feeding times. I realised how intelligent she was because she quickly worked out the moments when I would be slightly less in control, such as when I reached out for another syringe. She would wriggle and squirm and, on occasion, the surrounding furniture became sprayed with the smelly fish mixture. I always had to be one step ahead of her, but she often got the better of me. Every time I cured one problem, she would think of something else to do to be awkward, such as rolling on her side or wriggling backwards.

Several times I had to put Aqua back on the liquid-only diet for a few feeds as she had diarrhoea. She strongly objected to this but I knew that if it was not controlled it could be fatal. At last her system settled down and

What a joy it was when she gaped for food

she could have fish at every feed. When giving her the small pieces of herring it always paid to be firm, pushing each bit well over the tongue, as for the first few days she offered no help at all. Holding the mouth shut and stroking her throat encouraged her to swallow. She was always friendly and cuddly though – even after a bit of a tussle – and soon she wanted to suck again. She would suck her flippers when alone in the bath and suckled my hand when I offered it to her after a feed. She obviously drew comfort from this and it probably helped to bond her to me as her human mother. A young seal pup's fur is beautifully soft and silky, and it was a pleasure to stroke her while she lovingly nuzzled me under the chin.

What a joy it was when she began to gape for the food, and over the next few days she came to do everything for herself so I only had to pop it in her mouth. Once this hurdle had been overcome, I increased the size of the fish pieces until she could take half a herring at a time. Her care was much simpler now. She was drinking from her bowl and there was no more washing and sterilising of tubes and syringes. I no longer visited her at night and although she had a long way to go, I was not constantly peeking in her room to check on her as I had done at first. When she was sleeping deeply she was so still that I would watch to make sure she was still breathing, for until she got over her initial problems, I would worry that she could suddenly take a turn for the worse. A pup that is well usually sleeps on her side or her back, apart from after a large feed. A pup that constantly lies on its front is likely to have problems.

Aqua would have at least one bath each day with slightly warmed water and a little disinfectant. It was important to keep her clean, especially as she sucked her flippers, and she was also showered as necessary. Many people are surprised when told that seals do not have to be in water all the time. Young pups are even dried with a towel afterwards. However, as Aqua became older, she would frequently try to get into her washing-up bowl which was there for her to drink from, and once she had it tipped over she would climb on top as though it was a rock she had hauled out on. She would follow the shower as I washed round the bath, always trying to get her head underneath the flow of water, and she would do the same when the tap was on, diving beneath it and opening her mouth to catch the water. She enjoyed this so much that her next trick was to pull the shower off the taps. In the end I had to chase her up to the other end of the bath if I wanted to use the shower. She soon learnt to pull the plug out when having a bath. Realising that she wanted to play, I gave her children's toys which she pushed around with her nose as she swam. I had a scrubbing brush with a short handle with which I used to gently clean her. She would roll on her back and enjoy a tickle on her tummy with her fore-flippers up in the air and a look of pleasure on her face. It was very much like the soppy expression our Labrador Cliffy would have when he was scratched on his tum.

The next step with the feeding programme was to get her to pick up the pieces of fish for herself. She ignored the bits that I hopefully left in the dry bath, but when I left some in her water bowl, she put her head right in the water and began to feed. The only trouble was that she did not like bones, tails, guts or heads. When feeding her by hand, I had tried leaving little bits of bone or tail in the fish but, at the slightest hint of these, she would turn her head away disdainfully and refuse it. It was very frustrating. Eventually, I decided to give her everything filleted and hoped she would come to her senses later. However, it meant that I had a lot of discarded bits to dispose of. I would collect them over a few days and then take them down to the pier in a bucket and tip them into the sea where I knew the crabs and little fish would soon make a meal of them. Through the clear water I could see crabs of all sizes come out of their hidey-holes and make a beeline for the herring pieces. They were not the only ones to benefit from Aqua's fussy habit.

One calm evening, a seagull flew in and, after circling a few times, landed on the still, pale blue water. He bobbed along to pick up some pieces that were still floating and soon another gull appeared and did likewise. I withdrew to the edge of the woods and sat watching them for a while. These much-maligned birds looked quite beautiful to me in the setting of the quiet sun-filled bay. After this they often came to the pier for scraps.

I have tended several seabirds over the years and kept a herring gull in a shed with a half door for two months while his broken wing mended. He had been found on the road, one wing dragging at his side. I strapped Jonathan's injured wing to his body with bandages and sticky tape, and added antibiotic powder to his drinking-water each day. He did well on

From a rocky shore he flew strongly from my outstretched hands to freedom
(Chris Davies)

tinned dog and cat food and scraps. When I handled him I put a sock over his head which worked like magic to calm him down and protect me from his strong beak. I had done the same with our gander who had a broken wing a few years before. He is still going strong today. I desperately hoped that Jonathan was going to be able to fly again, as the quality of life for a flightless gull would be very poor. Besides, I didn't like the thought of him living in captivity for 20 years. After six weeks, I removed the bandages and took him up to the castle lawn for flying practice. It was a long haul, and on the first attempts he would just flap along never leaving the ground. Then, after a few days, he managed to rise a few feet into the air. At last I felt there was hope for him. At each practice session he improved and his once useless wing took him further and higher. The time came for his release and I carried him in the boat as we made for the islands. From a rocky shore he flew strongly from my outstretched hands to freedom.

Not all birds are so fortunate. Sometimes their bodies are so badly injured or full of infection that the vet has to recommend putting them down. Others are found dead on the shore or in shallow water at the sea edge having been fouled in discarded netting.

FEATHER AND WING

I've often watched the gulls in flight before,
seen them wing their way across the rocky shore,
seen their silver shadows fly,
'cross the ever changing sky of my island.

Feather and wing,
riding the wind,
feather and wing,
riding the wind,
ride, ride on the wind.

I've never seen a gull so still before,
he may ride above the stormy waves no more.
Touch his softly feathered wing,
he a fledgling from the spring of my island.

Feather and wing,
riding the wind,
feather and wing,
riding the wind,
he may fly no more.

I buried him above the sandy bay,
and I prayed that he should fly again one day,
lift his softly feathered wings,
ride the ever changing winds of his island.

Feather and wing,
riding the wind,
feather and wing,
riding the wind,
ride, ride on the wind.

One fine day, when Aqua had recovered from her early problems, I carried her down to the bay below the house. I had Peter and George there to help, for although she always seemed to want to stay close to me as her human mother, I didn't want to take any risks. They stood out a little way in the sea so they could recapture her if she seemed to be going out too far. Kicking off my wellies, I set her down on the warm sand and walked into the inviting shallows. Aqua followed at my heels like a well-trained dog. When very young, she had a funny one-sided wiggle as though she was not in control of her rear-end. I thought she might set off swimming once she felt the water of her natural home around her, but she stayed constantly close to my feet as I paddled up and down. When I knelt down in the water, she swam slowly round me, nuzzling my body from time to time. I went out a bit further and she climbed on to my tummy as I lay down. While I stroked her head and cuddled her like a baby, I knew she had no intention of leaving, for we had a bond that would not be broken until natural instinct told her it was time to leave her mother.

From then on Aqua became part of summer beach activity and joined picnics with the children, when I would take a bag of fish pieces for her snack. As she grew heavier, I would put her down at the bottom of the slope from where she flopped through the grass and tumbled down the rocks to the beach. At first she was quite happy to lie there with us as we had our picnic, but it wasn't long before the sea drew her like a magnet and, since by now I was confident she would not swim away, she was free to explore the shallows of the bay on her own. She would often look to make sure I was still close, though, and when she wanted a rest she would come to the edge of the beach, lying in a few inches of water but never coming out on to the dry sand.

She was so gentle and relaxed in the water that the children could play quite safely while she swam around nearby. I could take her anywhere I wanted by holding on to her fore-flippers and guiding her along. She learned to chase the fish pieces that I drew along in the water in front of her before I finally let go of the imaginary prey. In deeper water, I would roll her around on her side like a barrel, or hold her up against my chest from where she would dive with a splash, wiggling away for a few yards before turning back and zooming past my legs. She obviously loved these games and showed a great sense of fun. As she grew older we were amazed at her turn of speed; she was like a torpedo streaking smoothly and relentlessly through the water. I envied her ability to move with such ease from the world of air to that beneath the surface of the sea, as though the two were identical, her wide-open eyes never flinching with the contact of water. I would paddle at the edges of the bay, where the seabed was strewn with boulders and wrack lay in the shallows at low tide. With whiskers alert she explored amongst the half-submerged rocks, sometimes coming up for air with a mop of brown or green seaweed adorning her dark, glistening head.

I often played my violin in the sea while Aqua swam around me. One day I had been concentrating on my playing, when suddenly, I noticed a seal's black head by the isle in the middle of the bay. My first worried thought was that Aqua had gone out rather far by herself. Just as this head slipped beneath the surface, another head appeared a few yards inshore from the first, and I was shocked to realise that, in fact, this was Aqua. I was filled with foreboding that she would go off with the other seal and leave me, for she was not yet ready for the wild. By now both seals were somewhere underwater; with George on the beach and I feeling helpless in the sea, we held our breath. A seal surfaced slightly further out but, torpedoing faster than I had ever seen her, Aqua came streaking straight for me through the clear sea. She was all silly and excited and wanted to be close to me. It was obvious that she had met the wild seal underwater and it had given her such a fright that she came rushing back to mummy. I was intensely relieved that she had returned. I knew the visiting seal must have deliberately swum in because of Aqua's presence, having never seen wild seals coming so far into the bay before.

Sometimes our peacocks and peahens would troop down to the bay when they heard the family on the beach, looking for any scraps left over from

I went out a bit further and she climbed on to my tummy as I lay down
(Chris Davies)

the picnic. A pair of wild swans with their cygnets also learnt that the sound of children's voices and violin music meant a snack might be forthcoming. Although they were slightly apprehensive of Aqua's presence at first, seal and swans kept a respectful distance from one another and peace prevailed.

When I was on my own with Aqua, I enjoyed the freedom of bathing naked in the sea. It was a beautiful experience to be swimming free with a seal for a companion in the exquisite setting of this private, secluded bay. Aqua's antics, however, sometimes became a little difficult to handle. In her efforts to get on my back the claws on the end of her flippers would scratch me. Sometimes she would playfully and harmlessly nip at my heels, but once, she appeared to be taken over by some very basic instinct when she went straight for one of my nipples. Fortunately, I managed to get out of the way in time and laughing gently dissuaded her from trying again – but after that I always wore a bikini and often a T-shirt as well.

Aqua was brought in to be with the family when she had got over her infection and was stronger. By this time she had got used to the children's

From then on Aqua was part of summer beach activity

inevitable noisiness as sometimes I had to take one or both of them with me when I went to feed or shower Aqua. The dog appeared to happily accept her presence in the room, having already encountered her flopping around the courtyard and on the beach. Although she had a few accidents, I learned that she would relieve herself before I brought her into the room if I showered her first in the bath. It seems that water stimulates a seal to move its bowels.

Because she was wet, I used a towel to carry her out. She was very fond of her towels, scrabbling them up into a comfortable heap to lie on in the bath. Once, I hung the towel up on the fireguard and later on in the evening Aqua pulled it down with her teeth, rushed across the room with it and, after making it cosy with her fore-flippers, lay there happily.

Her behaviour in the living-room generally followed a pattern. When she first came in she would busily investigate everything within reach. Children's toy boxes had to be shoved out of the way because, especially in her early explorations, any container that was remotely seal-sized was got

Sometimes she would playfully and harmlessly nip at my heels

into. The wood basket held a particular fascination for her – possibly because it contained plastic bags of kindling and she had been fed little bits of fish from a plastic bag on the beach. She flopped into the low, open-shelved cupboard where videos and books would be flung far and wide by flippers and nose. After a good sniff around here and there, Aqua would start her really crazy antics which had us in fits of giggles. She would suddenly put on a burst of speed across the room, stop just as suddenly, then take off again in another direction, have a poke and a sniff at something, then off again.

Then there were the games with a little bit of paper or a feather. She would put her nose right up to it and blow hard through distended nostrils to make the object leap a few feet ahead of her. With an air of satisfaction, she would study it for a couple of seconds, reach it again with a series of little shuffles and then repeat the exercise. After doing this half a dozen times, she lost interest and settled down for a sleep.

Even in sleep Aqua was fascinating. There is a feeling of utter contentment that exudes from a snoozing seal. She was quite happy with myself and the dog, but any other person who entered her haul-out space received a warning hiss as she thumped the floor with a flipper several times in rapid succession.

She liked to lie next to Cliffy or with her head on my feet as I sat in the armchair by the fire. She would come in dark and glossy after her shower and, as she dried out, the beautiful patterns on her furry coat gradually appeared. Now and then she had a little scratch at her neck, gave a big yawn or curled her hind-flippers over each other. A seal's head and body are incredibly versatile – scrunching up short and fat or stretching out long and thin – and Aqua would change from one to the other as she lay half asleep with her eyes closed, occasionally spreading her hind-flippers out wide like two fans.

She watched television avidly, as did her successors. Music would set her off on a beautiful seal ballet, tail and flippers waving, supple neck stretching and weaving. The television, normally mounted halfway up the wall, was put on the floor for her convenience. She investigated the rear as if looking for a way in. George took a highly amusing video of her watching the animals on the *Really Wild Show*, stretching her neck out and tentatively touching the screen with her nose from time to time. She also had her own video of herself to watch – including a sequence in which I was playing the violin to her outside on the grass. Soon the screen and speaker became smeared with herring-scented snuffles.

Aqua was so inquisitive and fearless that I had to keep her out of the

Aqua and the rabbit got on fine. She would snuffle in the rabbit's ears and they would touch noses

George took a photograph of her showing off her slogan

She was truly bonded to me and no one else would do

way when I was hoovering, for she insisted on following its progress over the carpet. She never showed any fear of the dog, and Cliffy was completely tolerant of Aqua's little games. She would gently catch hold of his tail and pull it, and he would just look round. She would shuffle up behind him and blow in his ear, and although the ear twitched involuntarily, he ignored her.

Aqua often sunbathed in the courtyard, flopping into a shady spot when she became too hot. Hannah had a white rabbit which she used to push around in her doll's pram. Aqua and the rabbit got on fine when it was out for a run in the courtyard. She would snuffle in the rabbit's ears and they would touch noses. Aqua wrecked the children's paddling-pool but the episode gave us such a laugh that we didn't mind too much. She hauled herself over the side into the water one sunny afternoon and became very excited, hauling out the other side, getting in again, splashing through the water and flopping out again. She was in one of her very silly, fun moods, and every now and then she bit at the air-filled plastic. Gradually, the pool collapsed as the family did likewise in laughter.

She had her own pool in the courtyard – a deep white tank that we filled nearly to the top with water. To begin with, she would constantly look over the side and get upset if I went too far away. But after a few times she relaxed and played by herself quite happily, chasing her tail round and

Peter caught fish for her and even offered her a salmon

Loganair flew the fishy freight to the island for us

She looked fat and healthy, having by now developed a thick layer of blubber

I sat on the grass playing 'Islay Mist' on the violin

round, enjoying the deep water and the plastic balls I put in for her to play with. When a super quarry, which would undoubtedly have badly affected the lives of the seals, was proposed for Islay, 'Aqua Says No Super Quarry' was written on the side of her tank as her protest. George took a photograph of her showing off her slogan.

After a good swim in the sea or in the tank, she would settle down for a snooze. The living-room doors opened out into the courtyard and she would often choose to go into the house. Her sense of direction amazed us, for very soon she learned that the two routes from outside led along a corridor to her own bedroom where she would lie outside the door. Now that she had a routine of being taken out for various activities, she was quite happy to go back to her bath for a while. When it was time to come out again, she would help me pick her up, reaching toward me and pushing her hind-flippers against the bottom of the bath. Once I had to go to the mainland for one day and I left instructions with George and Peter to feed her and take her out for a swim in the courtyard tank. When I returned

OPPOSITE:
I gave her a last cuddle

130

The bat flew out and landed on my hand as I played the violin

they told me that she had eaten her food but when they came to take her out, she would not let them near her, hissing at them and giving the warning sign with her flipper. She was truly bonded to me and no one else would do. Although quite relaxed with family members, she would become very quiet and still if strangers were in the house. She knew right away if visitors were in the living-room, and would hang around at the door instead of coming in and making herself at home as usual.

The time came for her to move from the bath, which, by now, had to have a barricade to keep her in. She lived in the large blue tank that Don and Dot had first used for Ellen and, because she now weighed around four stone, for ease of handling, the children's slide was put by the tank. I used the steps as a ladder to carry her up and held her on a wide ledge while I got in the tank and lowered her in. To get her out, I placed her at the top of the slide and she slipped down by herself and flopped toward the back door to get into the house. The tank was between our bedroom and the steep slope to the bay. When I got up in the mornings, I would peek out of my upstairs window to see how she was, and she would stare up at me, having noticed my face right away. It was only at this stage that she at last began to eat the whole fish I left for her, although she would still take off the head and tail. I was usually able to buy fish and some of the Islay fishermen kindly contributed part of their catch. Sometimes Peter caught fish for her and even offered her a salmon, but this was at an early stage in her development and she just sniffed at it. When local supplies dried up, I had to fly fresh herring in from Glasgow. Val Johnston, an Animal Concern volunteer, toured Glasgow fishmongers buying up all their supplies of herring. Loganair flew the fishy freight to the island and Julie McNicol, one of their ground staff at Islay, made a special delivery to Kildalton. This unusual extension to their air ambulance service was fully featured in the Loganair in-flight magazine. Aqua looked very fat and healthy, having by now developed a thick layer of blubber to protect her from the cold sea.

One evening we had a visit from Eva Gunneberg, a researcher for the BBC's *Highway* with Harry Secombe. She was on Islay to collect material for a future programme when, on visiting the reception desk at the new swimming-pool, she saw a photograph of Ellen that Chris Davies had taken. We had given the pool some cassettes and song booklets to help raise funds for running expenses. George and I discussed with her the possibility of appearing in the programme and we looked forward to meeting Harry Secombe.

A television crew came to film a news item for the breakfast programme *Channel 4 Daily*. It was about renewed calls by fishing interests for a mass seal cull in Scotland. They filmed wild seals out in the sanctuary and John Robins was interviewed on one of the islands. He explained that there was an unofficial cull going on and the law should be changed to protect seals properly. They filmed Aqua in her courtyard tank and swimming in the sea with me while the children played in the water nearby. I told of my fears for her future – for she was soon to be released – saying that although she would be safe around Islay and Jura, young seals are known to travel long distances and she could be deliberately harmed elsewhere.

Gradually, Aqua grew more independent. I still brought her into the living-room in the evenings to watch television, but she was not so keen to be handled and I could see our special bond was loosening, She would still lie by me, happily waving her neck and tail as I sat out on the grass playing 'Islay Mist' on the violin. On one such occasion we had unusual company, for the evening before I had rescued a bat that was lying torpid. I brought it in, putting it in a warm box, hoping it would revive. I knew it was in danger of being killed by a rat or a feral cat if I had left it where it was. The next day the bat had not moved and in the afternoon I put the open box out in the warm sun as I played with Aqua. George was there to take some pictures of her. Whether it was the heat of the sun or my violin playing that woke it up I do not know, but, anyway, the bat flew out of the box and landed on my hand as I played the violin; my hand being the nearest thing to land on. After staying there a few minutes, it suddenly took off again and flew away through the trees.

Aqua weighed nearly 70lb by this time and she began to protest strongly when I brought her back from the sea. I knew the time had come for our parting and I met it with mixed feelings. I rejoiced that she had thrived and was fit to go back to the wild, but I also knew how much I would miss her. I had no idea if she would stay around Kildalton or disappear forever. For her sake, I had gone around smelling of herring for three months. She was to leave us when over 50lb heavier and our pockets were many hundreds of pounds lighter from paying veterinary fees, buying medicines and, of course, a lot of fish.

It was a calm day at the end of August when I took Aqua down to the bay with George. A few days before, she had stayed out in the deeper water for two hours but returned to the beach, obviously waiting to be taken home. This time I felt sure she would go for good. I took her deep into the water and gave her a last cuddle as she looked at me with her lovely dark eyes. She swam away from me, returned and swam past me several times. George was taking photographs which were to be used in press coverage of Aqua's release – linked with a warning not to handle a wild seal pup, but to get expert help if one was found that seemed to be in trouble. By now he had taken many different photographs of myself, Aqua and the children, and some of these appeared that December in a double-page colour feature in *Woman's Weekly*.

As Aqua swam up and down, a wild seal came in close and seemed to wait for her. Aqua disappeared underwater and then I saw two seal heads surface well beyond the bay. This time I felt sure she would not come back to me. I went to the bay later on in the evening and for several days afterwards – just in case she had become exhausted or unable to cope in some way. But the bay was empty, and my Aqua had returned to the wild.

EIGHT

Morag

As the warmth and calm of summer gave way to colder nights and autumn gales, I wondered how Aqua was getting on. Occasionally, she had been seen in the bay or further out to sea, sometimes on her own but more often with several other common seals. In fine weather, a group of them hauled out on the island in the centre of the bay. Aqua was completely wild now and I was pleased how she had integrated with the others – but she hadn't forgotten her unusual upbringing entirely. As George and I stood on the pier one calm evening, she appeared round the rocks with a couple of companions. Seeing us seemed to put her into one of her silly moods as she leapt over and over like a porpoise across the bay. She was such a delight to watch. We knew she must still be healthy if she was fit enough to perform these antics, but I still wondered how she would fare during her first winter months in the wild.

We had been right to let her go when we did, for if we had kept her much longer her dependence on us may have become too strong and her natural wild instincts may have been suppressed. From my bedroom window I looked down on the empty blue tank and across the seaweed-strewn beach to a grey and stormy sea. I would often scan the shore, believing that if Aqua was in trouble she would return to this bay for help. The television was back on the wall, but we had wonderful videos to watch which reminded us of our much-loved summer guest.

One Saturday morning, I had just got out of the bath when the phone rang. Hurriedly wrapping a towel around me, I answered the call in the bedroom. A young visitor to the island had been walking with her husband along the wild coastline on the other side of Islay. They had found a seal lying a little way up the beach in one of the coves of these exposed and cliff-lined shores. Judging from the blood she had seen at its tail-end, it appeared to be injured, but it was hard to tell the age or species of the seal from her description. She was calling from a telephone box and I arranged

135

to meet her on the small public road from where she would walk us to the spot. It was going to take us at least an hour to get ready and travel all the way over there. As it was a Saturday I contacted the vet, Paul, at home and arranged with him that I would phone him again when and if I had the seal at his surgery for him to examine.

I quickly got Hannah and Fifi ready for their unexpected outing and prepared boiled water, tubes, syringes and rehydration powders. I took a big old blanket and some towels. In my ignorance, I thought the seal might be like Aqua and that I could pick it up and carry it in a towel. The blanket was for it to lie on. Little did I know. We piled into the Range Rover and set off on the long journey. I didn't often get over to the other side of the island and was looking forward to visiting a part of Islay whose character so differs from that of Kildalton.

We drove round Loch Indaal and down the Rinns of Islay, turning off on to a lonely single-track road at Port Charlotte. Uninhabited but for a couple of farms, the view of bleak, rolling hills has only recently been broken by commercial forestry enterprise. We met Morag Buchanan and her husband, Jamie, by the ruins of the chapel dedicated to St Kiaran, who came from Ireland to Kintyre in AD 536 in the days of St Columba. The chapel lies close to the head of Kilchiaran Bay where the saint is believed to have landed on his way to Iona.

It was only a little further along this coast that the Seal Action Group had earlier performed a rather different rescue. In an area where pounding waves batter the high cliffs, a very young lamb had somehow managed to become stranded on a ledge in a steep gully; probably chased there by dogs. The slowly starving lamb was stuck on a rock which was cut off by the sea entering the inlet and, if she could not be rescued, the only other option would have been to shoot her to prevent further suffering. Eric Bignal, a farmer and a conservation expert within the Seal Action Group, felt that, if at all possible, the lamb should be given a chance to live.

The coastguard considered any operation to save the lamb too dangerous. They couldn't guarantee the safety of anyone attempting to go down the cliff on a rope, and to approach the rocks from the sea would endanger the lifeboat. Eric called on fellow members of the Seal Action Group for help. Alan Gurney, a yacht-designer and a lecturer on Antarctica living in the converted kennels on the side of the Fairy Hill, lent us his small inflatable.

After picking up the boat from Cnoc Bay, we drove over to the Rinns. It was transported on top of a Land Rover as far as possible, then it had to be carried a long way over rough ground. The boat was heavy, but a lighter craft would not have been safe in the rough seas entering the gully. We reached the top of the cliffs and could see the lamb looking small and pathetic far below. The boat then had to be carried to the foot of the rocks over piles of treacherously slippery boulders. A long rope was tied to it for safety before Eric and his son, Robin, ventured out into the water.

Beyond the inlet the ocean waves roared as the little craft was eased toward the lamb on the ledge near the mouth of the gully. Robin picked up the freezing-cold, wet lamb and with the help of the rope, the rescue boat was returned to the safety of the shore. Finally, the very difficult haul of the boat back up the gully was completed and it was placed on the Land Rover. The lamb, now named Selkie, was taken to the farm and nursed to a full recovery. The unusual episode was reported on the front page of the *Oban Times*, illustrated with a photograph of Robin and the lamb which was taken by George at the rescue site.

Mindful of the difficult rescue of the lamb, I took the towel and, entering the gate, began the coastal trek on grassy slopes high above the cliffs of this majestic bay. Heavy rain came in squally showers as we looked down on a rough and foamy sea, and we were reminded by the assaulting gusts of fierce wind that there was nothing but miles of wild Atlantic ocean between us and the coast of America.

Following a rough farm track, we eventually arrived at the head of a

Robin picked up the freezing-cold, wet lamb

narrow, sheltered bay, bordered on either side by rugged lines of rock which stretched well into the sea. We were told that the seal was on the beach, but at first I could not see its camouflaged rotund form – tiny against the gigantic dimensions of this magnificent scenery. Having spotted it, and thankful it was still there, we negotiated the boulders and moved down on to the beach. As we drew closer, I realised from its parallel nostrils and slightly Roman nose that this seal was a completely different kettle of fish to Aqua. I had known there was a possibility on this side of the island that the seal would be a grey pup, as Morag had given the impression that it was not very long. The grey or Atlantic seal pups mainly in the autumn on exposed and remote shorelines. The exact timing varies between breeding sites and, during the aftermath of the Shetland oil disaster in early January 1993, I heard from wildlife rescuers of their surprise at seeing baby whitecoats in the region. I did not know exactly when grey seals had their pups on Islay.

I had only just managed to lift Aqua by myself at 70lb, and she, of course, was friendly. There was no way I was going to be able to pick this one up. Although about the same length as Aqua, she looked nearer 90lb, and as I drew closer to examine her hind-flipper which was injured, she turned quickly. Darting her head out on outstretched neck, she took a bite at my leg, but luckily, only my trousers were damaged. I had ascertained that she was a female from her creamy-white coat covered in dark splodges, and a further examination of her tail-end would confirm this. The injured hind-flipper showed an exposed bone and was bleeding. Without the use of her method of propulsion, she had no hope of surviving in the sea.

She would have been born weighing only about 30lb, but during the intense lactation period of less than three weeks, spent on land in her fluffy white coat, she had more than tripled her weight to around 100lb. As she began shedding her baby covering to reveal the beautiful markings of her permanent coat, her mother would have left her. At this stage the young seals have to learn to feed and fend for themselves in the rough waters of approaching winter, and here was a pup that had somehow been injured at this critical and vulnerable time. However, I had been expecting either a common seal pup of a few months old, or a younger grey which had become separated from its mother and, therefore, had not yet grown to this enormous size. And here we were on a remote shore with nothing but a pathetic little towel to pick her up in!

As George and Morag set off in search of help, I stretched the towel between my hands and used it to dissuade the seal from returning to the sea. Jamie helped by keeping Hannah and Fifi amused as they played with shells and pebbles on the beach. We were all getting pretty cold and wet

I used the towel stretched between my hands to dissuade the seal from returning to the sea

as we waited in the rain. At least half an hour later, we heard a vehicle approaching along the rough grassy track. A sturdy Land Rover appeared and stopped at the head of the bay. The farmer from Kilchiaran, Neil McLennan, and his son had kindly come to the rescue.

George had brought the old blanket and as the others held it on the ground by the corners I persuaded the seal to shuffle on to it. She was then carried in the makeshift sling all the way up to the Land Rover and heaved as gently as possible into the back. Here I sat with her, the blanket, still partly in place, keeping her from attempting to bite again, as we slowly wound our way along the precarious route by the cliffs back to our starting point at Kilchiaran Chapel. Here the seal was transferred to our Range Rover and after exchanging addresses with the Buchanans, we headed off to the vet at Bridgend.

On receiving my phone call, Paul came over to the surgery immediately.

I prepared the bath for the seal, now named Morag, with bigger barricades and an extra chair for support

He examined the seal in the back of the Range Rover but was not very hopeful as the swollen hind-flipper had a broken bone and there was a danger that the open wound would give rise to serious infection. We went into the surgery and while we warmed ourselves with hot cups of coffee, Paul telephoned Orkney and Pieterburen in Holland for advice. It was recommended that he should contact vet Ian Robinson at the Norfolk RSPCA Wildlife Hospital. Ian, known as the 'Birdman of Basra' after his work in rehabilitating birds oiled in the Gulf war, was considered to have the expertise necessary to tackle this particular problem. It was vital to control the infection in the hot and swollen flipper. We were to take the seal home with us and administer two types of antibiotic, as well as applying antibiotic cream externally to the wound after cleaning it out daily with a special fizzy fluid. Paul planned a visit in a few days' time to see how she was progressing.

We drove on to Bowmore where I visited my mother who was a patient in the cottage hospital. I explained to her that I would not be making my usual visit that evening as I had to look after an injured seal. She was amused that we had the animal lying in the back of the Range Rover in the hospital carpark. As we made our way home, we were amazed how the seal seemed to relax and looked out of the window at the passing scenery.

When we arrived home, I prepared the bath for the seal, now named Morag, with bigger barricades and an extra chair for support as I knew her weight made these precautions necessary. I had not yet attempted to give her rehydration fluid by tube because I knew it was not urgent in her particular case, and I would need a lot of help to control her. I called on Big Donny – Donald McKinnon from Lagavulin – who had helped move the grand piano. When he arrived, we carried the seal in the blanket on to the living-room floor. Here Donny sat firmly on her back, facing her head. At first she bucked and struggled, nearly getting the better of the big man, but she calmed down once Donny got the hang of the technique. I administered rehydration fluid, laced with vitamins and the all-important antibiotics, down a stomach tube. I also gave her Arnica, a homoeopathic medicine, to calm her and aid in the healing process.

The family looked on in amusement as this little roly-poly barrel of a seal showed how strong and full of life she was despite her injury. Luckily, no one got bitten and Morag was transferred to the bath where her wound was cleaned out with the shower. I had to frequently bathe the hot swelling with cold water. I managed to attend to her flipper by keeping the deck scrubber between her biting end and my hand. Due to the size of her blubbery body, she really did not have much room to move about in the bath. For the time being, though, it didn't matter. She was safe and in with a chance of survival, and perhaps her confinement was a good thing since she was forced to rest her injured flipper.

On the second day, after seeking advice from Maureen Bain on Orkney, I started to feed Morag with whole herring. I still had some left over in the freezer from Aqua's stay. I had to get into the bath with her as moving this seal on to the floor every time was impracticable. I knew she was bound to try and bite me out of fear so I put on an extra pair of trousers and sweatshirt as well as my long, thick leather gloves. After removing the barricade, I got the blanket ready and put a couple of herring within easy reach. Morag looked at me soulfully. She was so beautiful with her creamy coat that shone silvered as it caught the light. She wore a little smile on her face below which rolls of puppy-fat gave the impression of a double chin. A spray of strong, stout whiskers grew from either side of her nose and the

With help, I was able to take her out into the courtyard each day

more parallel nostrils peculiar to her species opened and closed with her breathing. The dark fur on her head was velvet soft as were the eyes that watched my every move. Like a cluster of antennae, more whiskers sprouted above the brow of each searching eye. Reminding myself that, despite her size, this was but a baby of three weeks old, I threw the blanket over her and entered the bath at her tail-end.

Morag lay still as she was completely covered. Straddling her and controlling her movements with my thighs, I gently removed the blanket from her face, holding her head with both hands. She was quiet now and as I spoke to her gently I picked up a fish. I was surprised how easily I opened her mouth from the corner and introduced the herring, head first. What a big throat she seemed to have compared with the little common seal pups I was used to. She swallowed the fish whole and from then on I was able to give her the medicines tucked inside her food.

Over the space of just a few days, Morag became friendly and playful as she grew used to me feeding and nursing her and keeping her bath clean. It was an astonishingly rapid change from the frightened, struggling creature we had first encountered. She craved affection and would actually try to cuddle up to me, putting her fore-flippers on my knees and enjoying my caresses. She lost all of her initial aggression and I felt quite safe in the bath

We met them with Morag in the back of the Range Rover

143

with her, but I underestimated her playful behaviour. She loved the shower and made it obvious that if it was on but not directed at her, she was going to kick up a fuss. On one occasion, when I thought I had at last persuaded her to stay at the tap end while I cleaned round the bath, she lurched forward and took a nip at my knee. I understood it was only in play with no hint of real aggression, but I had learned to my cost the difference in character between the grey and common seal pups who I had never known to behave quite like this. Upon investigation, I discovered that in that split second Morag had given me a nasty bite, with three puncture holes and a surrounding bruised area. Luckily, it healed with no sign of the unpleasant infections which I knew were possible with seal bites. After that I wore even more layers of clothing when working next to her and looked something like the Michelin Man, but she never bit me again. With help, I was able to take her out into the courtyard each day for a little exercise and was impressed by her turn of speed as she lolloped along, hens and ducks scattering in alarm from her path.

A few days after her arrival, Paul came to visit Morag to assess her progress. He was delighted and amazed at the improved state of her wound. It was clean and the swelling had reduced, although she could not spread the flipper out in the normal way. Paul contacted Ian Robinson again and it seemed that the best option for Morag would be to go to Norfolk and take advantage of the wonderful new facilities that the RSPCA had built there specially for seals. The hospital was able to provide on-the-premises X-ray and pathology facilities as well as the expertise required to anaesthetise a seal if an operation was needed. Seals present a unique problem in that they do not continue to breathe naturally under anaesthetic. If Morag was to be returned to a normal life in the wild, it was vital that her hind-flipper became fully functional again. The hospital had large pools in which she could exercise and regain her strength.

With the help of John Robins of Animal Concern, Morag's travel arrangements were organised. Dougie Walker, a veterinary nurse at the hospital, drove all the way from Norfolk to the ferry terminal at Kennacraig, conveniently stopping off for a family wedding on the Scottish mainland. He and another nurse took the ferry to Port Ellen where we met them with Morag in the back of the Range Rover. As there was no room on the ferry for Dougie's vehicle, the ferry company Caledonian MacBrayne had arranged that the seal could be put in a box in the back of a lorry that was making the crossing. Dougie had brought a large plastic box, specially designed for animal transportation, and Morag was transferred into this with no trouble.

Hannah was at school but Fifi joined George and myself on the two-and-a-half-hour journey to Kennacraig. I spent most of the trip sitting with Dougie and the other nurse in the back of the lorry next to Morag in her box. She soon settled down and we chatted about looking after seals in general and of how Morag would be cared for on the long journey south. She was to be checked periodically, her temperature monitored and her eyes and flippers sprayed as necessary.

When the ferry docked and we had disembarked, Morag was transferred from the lorry to Dougie's car. Her box was opened for a short time and I gave her a last hug as a reporter from the *Oban Times* took a photograph of us which later appeared in their newspaper. It was funny to think that I was able to do this quite happily, whereas just six days earlier I had called on the help of a big, strong man to control her – Donny had said he'd thought 'she would bite the hand off you'. Morag looked so adorable now and I knew I would miss her, but I was looking forward to her returning fit and well. She would be grown up and changed, no longer a playful pup, and I would never have those tender cuddles with her again. All being well, she would return to Islay in the spring, but at Norfolk RSPCA hospital human contact is kept to a minimum for fear of the seals losing their wild instincts. A seal that is no longer wary of man would be in danger when returned to the sea.

While Morag was away the *Highway* television crew came to Islay to record the programme for which researcher Eva Gunneberg had visisted us in the summer. A cameraman with diving equipment went out to the sanctuary to take pictures of seals swimming underwater. Before coming to the farmhouse, Harry Secombe was filmed visiting the Kildalton Cross. As we sat round a table in the courtyard, he interviewed George and myself about the wildlife on Kildalton. George spoke of the beauty and power of God which is revealed in His creation in this wild, unspoilt place.

It was late afternoon before we headed out from Cnoc Bay to the seal sanctuary with the film crew. I sat on a small islet surrounded by the sea and played 'Islay Mist' over and over again while the cameras filmed me from many angles, including a shot where the cameraman was circling around the islet in a boat. It was a fine evening but the October air was cold and my fingers grew numb as I played. Above the splendour of the An Lanndaidh hill range billowed tenebrous grey clouds, but the clear sky overhead gleamed. From my rocky stage the crisp autumn air bore the strains of 'Islay Mist' over the dark, glassy sea to the Fairy Hill and the string of Seal Islands. When I returned to the boat my hands were so cold that I dropped my violin, fortunately without damage. The end result of the filming was worth it though, for combined with the underwater shots of seals, the scene was ethereal and evocative.

Above the splendour of the An Lanndaidh hill range billowed tenebrous grey clouds,
but the clear sky overhead gleamed

We telephoned the RSPCA periodically to see how Morag was getting
on. The X-ray showed that a small piece of bone was missing, but the
wound was healing and the limb was growing stronger despite this. On one
of his trips down south, George visited the Norfolk Wildife Hospital with
a poet and scriptwriter friend, Jim Vollmar. Jim had been to Kildalton and
written a *Vegetarian Magazine* article while sharing the spare bedroom with
Aqua. They were pleased to find Morag looking healthy and contented in
this Hilton of seal sanctuaries. She had been one of the first seals to benefit
from the new facilities and the large outdoor pool was still in the process
of being built during her stay. However, it was ready in time for her to
have plenty of exercise with the newly functional flipper before she was
returned to Islay for release. In seal terms, Morag had won the National
Lottery. She had been lucky to have been found in that remote and wild
place. Morag Buchanan wrote to us and explained how she felt that in some
strange way she had been led to that spot. Further to this, a brand new
million-pound hospital with special facilities for seals had just been opened

Eric Bignal and John Robins helped carry the heavy wooden box down to the shore

for Morag's benefit. Now she was to make the journey back to Islay in the RSPCA hospital's latest acquisition – a shiny, dark-blue Land Rover. It was important that she was released here, for in the Wash the seals develop different immunities and are genetically dissimilar.

One beautiful May morning George and I took the children down to the ferry terminal at Port Ellen to await the return of Morag. We were all very excited about seeing her again and wondered what she would look like after six months. We watched as the ferry slowly made its way into Port Ellen bay and docked at the pier. As the ramp was lowered the Land Rover came into view. Ian Robinson and Dougie Walker followed our car up the road to Kildalton, and in the courtyard I had my first glimpse of Morag as she was checked over and showered in the very large wooden crate in which she had travelled. She looked magnificent. Morag was noticeably larger and there was plenty of blubber on her, but she still had that appealing, dignified face that I had grown to love so many months before.

We proceeded to Cnoc Bay where a small audience had gathered. Eric Bignal and John Robins helped carry the heavy wooden box down to the shore. My mother, who was now living in the Dower House, sat in a deckchair in the warm sunshine with her carer, Jane McLeod. Although brought up in Islay, Jane had lived in Glasgow for many years and had only

Then she was in the water, head and back glistening in the bright sunlight

recently returned to the island. She loved the wildlife on Kildalton and was thrilled to see the seals that came into the bay below the house.

George set up the video camera and prepared to take photographs. Morag's moment of freedom had arrived. Ian and Dougie opened the box and she looked out on to the few yards of sand and the spacious sea before her. Slowly she ventured out and peered around before making for the water with a smooth undulating movement of her chubby but sleek body. She stopped now and then for another look around, then she was in the water, head and back glistening in the bright sunlight. As she moved in deeper and began to swim, Ian and Dougie followed her progress with binoculars. She swam to the shallow side of the bay and spent some 20 minutes there, as if taking stock of her surroundings, before heading off into deeper water and disappearing from view.

Don Bowness and George took Ian and Dougie out in the Seal Rescue boat, but Morag was not seen again that day. The commons were plentiful among the Seal Islands and Ian commented on the wide range of colouring they displayed and the good age-structure of the colony. The next day we took Ian and Dougie to Kilchiaran to show them the bay from which Morag had been rescued. While following the cliffs round Kilchiaran Bay, we observed a common seal playing in the large waves that rolled in from the

Jonah had never seen a whale close up before and was thrilled to have his
photograph taken as he stood on this leviathan

Atlantic. This is something we do not see on this side of the island for our
shores are so sheltered. On returning to the public road, a Land Rover
pulled up. It was Neil MacLellan who had helped us rescue Morag and we
were able to tell him of her successful release the day before.

We went on to visit Eric and from a distance watched a chough as it
gathered food in fields near Saligo Bay and returned to its mate and family
of nestlings. We marvelled at the rugged splendour of the wild Atlantic
coastline with the surf crashing in on the rocks, which has even been known
to occasionally cast up a dead whale. It was here that Hannah and Fifi had
their first encounter with one of these enormous sea mammals. It was a
minke and seemed huge compared to them. But it was small compared to
the sperm whale – a giant of approaching 50 feet – which was first
discovered lying on a rocky shore. Later, a high and stormy tide
permanently beached this giant even further inland. George took the
children to see this monster of the deep and it is with sadness that I now
recall how he also took a long-standing friend called Jonah to see the whale.
Jonah had never seen a whale close up before and was thrilled to have his
photograph taken for the national press as he stood on this leviathan. Sadly,
a few days later Jonah had a fatal accident and so the story of his meeting
with the whale was not published; instead George gave the photographs to

his relatives. On the way over they had played my album and, in particular, 'Saligo Bay' as they were going to visit this beautiful western coast of haunting grandeur and resplendent sunsets.

SALIGO BAY

Strolling, strolling the shore as the sun goes down,
you're standing there.
Your love, love comes to me as the seabirds cry,
it's yesterday.

Following paths we once both knew,
I trace our love in the sands as before;
you loved me then,
we were in love.

Strolling, strolling the shore as the sun goes down,
you're standing there.
Your love, love comes to me as the seabirds cry,
it's yesterday.

Where are you now as I wander alone,
knowing love dies and never returns? Oh, my love,
come back to me,
I long for you.

Strolling, strolling the shore as the sun goes down,
you're standing there.
Your love, love comes to me as the seabirds cry,
it's yesterday.

I feel our love I once knew,
time washes our kisses away and I cry,
I cry but our love remains in the sands,
I cry but our love remains in the sands,
forevermore, forevermore.

NINE

Uisge Na Beatha

It was early June and the days were calm and hot. Morag had been seen several times in the vicinity of the seal sanctuary, easily recognisable on land by the bright orange identification-tag that the RSPCA had put on one of her hind-flippers. Ardbeg distillery brewer Ian Hallam had arranged with George to take a trip out to the Seal Islands with his wife, Antoinette, and their little daughter, Jennifer. Don Bowness picked them up at Ardbeg pier in the Seal Rescue boat early one evening. Ardbeg consists of a farm, a few houses and, of course, the distillery buildings. With their distinctive pagoda-shaped roofs, they enhance the picturesque view from Ardbeg looking over

With their distinctive pagoda-shaped roofs, they enhance the picturesque view from Ardbeg, looking over to the tidal island of Imersay

151

to the tidal island of Imersay with its many ideal haul-out sites for seals on the surrounding rocks and skerries.

The three small settlements along the five miles of winding road between Port Ellen and Kildalton are all distillery based. In the village of Port Ellen, whisky is no longer distilled but barley is still processed to provide malt for all the distilleries on the island. Between Port Ellen and Lagavulin lies Laphroaig distillery, recently given the Royal Warrant by Prince Charles. Each island malt has its own distinctive flavour, produced with pure water from the Islay hills and local peat subtly flavours the drying malt at the kilning stage.

Our pony, Misty, was commandeered on several occasions to pose as the white horse for Lagavulin distillery during promotional receptions for foreign business guests. He had to be thoroughly washed until he was worthy of the role. Our friend, Alastair Robertson, was manager there at the time and I wrote a song for him and sang it in several languages at these enjoyable functions. Lagavulin distillery overlooks the Lords of the Isles stronghold, Dunyveg Castle, the ruins of which stand on a promontory at the mouth of the bay. In the year 1314, 1,000 Islaymen embarked from here in their little ships to fight at Bannockburn alongside Robert the Bruce. In those days, too, the timeless eyes of the seals would have watched as man performed for the history books.

LITTLE WHITE HORSE

Little white horse, trotting by the seashore,
did you see the ships sailing out from the bay?
Little white horse, standing by the castle,
I feel the past as though it happened today.

White horse, I'll take you anywhere,
where we can be free and far from care,
the hills are calling, the seabirds cry,
we'll fly with the wind, white horse and I.

Little white horse, trotting by the seashore,
feeling the wind blow from far away lands.
Little white horse, standing by the castle,
watching the waves turning rock into sands.

Little white horse, trotting by the seashore,
did you see the ships sailing out from the bay?
Little white horse, standing by the castle,
watching for your friends coming in on the waves.

White horse, I'll take you anywhere,
where we can be free and far from care,
The hills are calling, the seabirds cry,
we'll fly with the wind, white horse and I.

On one occasion George and I were invited to a reception at Lagavulin at which Olympic gold medallist Alan Wells was guest of honour. Both wearing their running gear, George chatted to the sprinter about the difference between their two sports and told him of how he had come first in the recent Bens of Islay Race. For safety reasons, this race – over 20 miles of wild terrain – is only open to members of Kildalton Athletic Club and with the late withdrawal of Roland, George had, in fact, been the only starter. To show he had completed the course, he had placed a miniature bottle of White Horse whisky, signed by Alastair, at the top of each of the seven bens. Alan Wells ran for under ten seconds with hundreds of millions of people watching, while George ran for over four hours with only the red deer and eagle observing him.

On another occasion we were invited to the distillery for the first official tasting of Bell's Islander whisky. On the bottle was a label showing a map of Islay with a legend reading, 'The standing stone above Cnoc Bay on Islay is said to mark the enchanted grave of Ile, a Danish Princess who drowned while bathing.' George was incensed. Apart from there being not one, but two standing stones on the side of the Fairy Hill, there was no way the princess was having a bathe. A headline appeared on the front page of the *Glasgow Herald*, 'Message on a bottle upsets Fairy Queen'. The article went on to say that a whisky firm had become the latest victim of the Queen of the Fairies whose hobby was making mischief, usually at the expense of outsiders. According to the Fairy Queen's disciple on earth, the distillers had got the legend wrong on the label of their newly launched Islander whisky.

On that June evening when Ian Hallam and his family had embarked from the small pier below Ardbeg distillery, they headed out to sea past the mouth of Seal Bay and the distillery's commercial pier. Here, in days long past, Clydesdale horses waited patiently with their drays while puffers with names such as *Moonlight* and *Warlight*, *Lady Isle* and *Inchcolm* were unloaded.

153

Seals would come around the pier and were fed fish by the friendly sailors. The wrecks of two puffers lie in these waters after hitting rocks as they moved out from the pier; victims, it is said, of 'a dram too far'.

The Seal Rescue boat passed the peninsula of Ardimersay and crossed the open waters to the seal sanctuary islands of Eilean Bhride and Outram. After a trip around the islands viewing the seals, George asked Don to take the boat to the pier below our farmhouse so that I could join them on this beautiful and serene summer evening. He knew I wanted to have a look around for Morag.

As we moved slowly among the rocks of Outram, a young, female grey seal appeared some 20 yards from the boat. She dived and rose again several times but kept her distance, and I could not be absolutely certain it was Morag. We were on our way home and dusk was falling when a slight movement made George look high up on the rocks of one of the small islands. He asked Don to slow the boat and move in closer. A lone, light-coloured pup peered around a boulder at the approaching boat-load of people. His dark eyes looked huge set against the tiny face and pitifully skinny body.

Convinced he had been abandoned, we carefully moved into the side of the rocks as I donned a thick pair of gloves. In spite of his small size he could have inflicted a nasty bite. I climbed out on to the rocks and as I neared him his plight became obvious. He didn't even bother to move, and he was so emaciated that the bones which are usually covered in blubber at the tail-end of the body stuck out. Terribly dehydrated, the dark wet 'mask' that surrounds the eyes of a healthy seal was completely absent. He must have been lying there high up on the grass for some days because ticks had taken the opportunity to infest his head and neck. He didn't offer the slightest objection as I picked up his feather-light body to take him home. Another day in the hot sun would have been the end of him.

On our arrival home, I began the rehydrating and feeding programme. His temperature was low so I used a heat lamp in the bathroom and gave him a comfortable, warm bed of old towels. I began a course of antibiotic treatment to stave off any possible infection. A few days later, our vet, Paul, was visiting on another matter and I took him in to see the seal. He looked a bit shocked but just said, 'He is very small, isn't he?' I knew he did not hold out much hope for this particular pup.

We called him Uisge na Beatha, literally, 'water of life' – the Gaelic name for whisky. Roughly pronounced 'ooshga na bear', it seemed an appropriate name since his rescue boat left from Ardbeg distillery pier and had the brewer among its passengers. Despite his appalling condition on arrival, Uisge grew gradually stronger and was gaping for fish pieces by the time – a little over a week later – another abandoned pup was brought in by George and Peter. She was in better condition than Uisge, but was still in need of urgent attention. Peter was sharing Uisge's room and so the en suite bathroom of a spare bedroom upstairs was prepared for the new seal. She was named Sophie as she was found on the birthday of George's young granddaughter.

Until the two pups were stronger and healthier, they were kept apart. They both made good progress and I took them into the courtyard, the living-room and the sea separately. As with Aqua, they bonded to me as though I was their mother and wanted to follow me everywhere. But this time, after my experience with Aqua, I deliberately had George junior and Peter handle them occasionally so that they could help me if the need arose. The two pups soon learned to suckle my hand – an activity which I gave time to after a feed as it obviously gave them comfort.

When both seals were obviously thriving, they were introduced. At first unsure of each other, they gave hisses and flipper-thumping warning signals, keeping a few feet between them as they hauled out in front of the television. But soon, after tentative nosing and more time spent together, they became relaxed and happy to be close. The two would cavort in the deep courtyard tank, chasing each other's tails round and round, twisting and turning in play.

Ian toasted Uisge's health with a glass of whisky

When the seals were very small, the baths did not need any barricades which made it easier to attend to them. When Uisge or Sophie heard me coming they would wake up and, as I walked in, I would see a beautiful little face thrust over the edge of the bath. I could tell if they had been sleeping as the mask caused by fluid from the eyes takes a few minutes to show after they have awoken. Uisge had a small red lump on his back near his neck when we rescued him. It grew steadily larger and then turned white on the surface. It didn't seem to bother him and the vet thought it was a boil that would disappear in time. When Uisge was stronger, Ian Hallam came with his family to see him. Both pups were perfectly safe to be with when they were small and little Jennifer was able to stroke Uisge's velvet-soft head as he lay on Antoinette's lap. We went for a walk down to the pier taking the pup with us and here Ian toasted Uisge's health with a glass of whisky.

Once the two pups were happy in each other's company and were both eating from my hand, I looked after them together as though they were twins. They were only separated at night and for rest periods during the day – it was I who needed the rest! The day would start with bathing the seals. From time to time I weighed them, standing on bathroom scales with a seal in my arms and subtracting my weight from the total. Then, if the weather was good, they were brought outside to be fed together, each with a bowl of carefully weighed fish pieces. It was a two-handed job to keep them both happy. Meanwhile, the courtyard tank was filling up for a morning swim. After their playtime in the tank, they would have a snooze or we would go for a little walk up the drive. Hannah and I would play the violin to them for a while out in the sunshine. If the weather wasn't good, Uisge and Sophie came into the living-room and lay by the television, sometimes watching a video such as Disney's *The Little Mermaid*. As they dried out, the beautiful colourings on their coats would appear. Uisge had a particularly attractive silvery pelage, while Sophie was generally darker and had many more spots.

But Uisge still had that one big spot which was now giving cause for concern. I bathed it with a mild disinfectant as it became pussy, but it did not develop like a proper boil. I phoned Ian Robinson at the RSPCA Wildlife Hospital in Norfolk and described the 'boil'. Immediately, he told me that the pup had seal pox. It would pass in time but was caused by a virus and it was too late to prevent Sophie catching it, too, as she had spent so much time close to Uisge. Sure enough, after a few weeks she became out of sorts for a few days and half a dozen small bumps appeared on her body which developed into pustules similar to Uisge's spot. Although Sophie had quite a few, they were all much smaller than Uisge's.

I also rang Ian Robinson for medical advice on another matter – but this time it was for myself. When teaching Uisge to feed for himself, he had been particularly eager and occasionally I had sustained very tiny cuts on my fingers from his sharp little teeth. I wasn't too worried and bathed them in disinfectant. But the cuts became inflamed and my fingers were swollen and very sore. It felt a bit like chilblains except that the pain didn't go away. Touching anything was agony and life became very difficult. I heard horror stories about people losing fingers or having to spend months in hospital after not being properly treated in the early stages of this condition known as 'seal finger'. The antibiotics the doctor had prescribed were useless, but after speaking to Ian, I was given terramycin which gradually cured the condition. The doctor had to lance a pus-filled boil that developed on my thumb. I had learned my lesson the hard way. Although I hadn't caught 'seal finger' from a deliberate bite, it was essential to wear protective gloves at that awkward stage when the pup is learning how to swallow the fish for himself, but still needs a little help.

On fine afternoons, the two seal pups would join the children as they played on the beach. If there was nobody about to help me, I took the seals together in the empty courtyard tank on a trailer pulled by the ATV bike. We had to take the long route through the field that slopes to the bay. As

They were brought outside to be fed together

We would go for a little walk up the drive

Hannah and I would play the violin to them for a while out in the sunshine

When set down on the beach, they flopped straight down into the shallows

with Aqua, when these pups were very young they were quite happy to stay on the sand and would only enter the water when I did, but as they grew older the sea was an immediate attraction. When set down on the beach, they flopped straight into the shallows themselves, but once water borne, would turn around to check I was still nearby. The tank was useful when the time came to go home as I could contain one seal in it while I fetched the other from the sea. Sometimes they would avoid capture, wanting to stay in the water longer, teasing me by swimming always just out of reach, eyeing me like naughty children. I soon found the answer was to grab a hind-flipper at an opportune moment.

I took fish with me and fed them together in the water as these seaside afternoons would include a mealtime. Uisge was fascinated by my bandaged thumb. Perhaps he was feeling guilty. He was always the more adventurous of the two and I often wondered if he had lost his mother because of his tendency to go further away from me than the other pups I had cared for.

Even when the two seals grew heavy, I could handle them together in the sea with the water buoying up their plump bodies

If he wandered too deep on his own, I would bring him back for safety. Even when the two seals grew heavy, I could handle them together in the sea with the water buoying up their plump bodies. In a few feet of water the two would play chasing games as they had in the courtyard tank. But here they frolicked amongst the seaweed, sand and rock, with clear, light-filled water reflecting blues from the sky and greens from the foliage that edges the bay. As they played joyfully together, I wondered if their puphood friendship would continue into adult life.

As if having two real seal pups sharing their lives wasn't enough, Hannah and Fifi had two plastic blow-up seals to join them at bath time. They had their own equipment to play with too, pretending to tube-feed their toy seals with old syringes I gave them. It was a big thrill for them when we heard from the BBC TV children's programme, the *Really Wild Show*, that they were interested in doing a feature on bringing up seal pups in a family. Although the finished article would only take ten minutes of a programme,

Hannah and Fifi had two plastic blow-up seals to join them at bath time

161

Hannah and Fifi sat with Michaela and myself at the courtyard bench-table as we cut up and weighed fish

the production team, including presenter Michaela Strachan, were here for the best part of a week.

They took a break to visit Eric Bignal on the western side of the island to do a feature on the chough. Eric studies the last remaining population of these birds on Islay, which shares the rugged habitat with the eagle and peregrine falcon and breeds in the area of his farm. There are less than 300 breeding pairs of this species left in Britain and it has completely disappeared from Cornwall which traditionally has the chough as its emblem. Prince Charles, the Duke of Cornwall, is in support of a recent programme to reintroduce this endangered member of the crow family. With its shiny black plumage, red legs and curved red beak, the chough wheels, dives and soars in spectacular flight along coastal cliffs. In legend it is said to bear the spirit of King Arthur. Michaela explained how rare these birds were but that 80 breeding pairs – 90 per cent of the Scottish population – live in Islay. She interviewed Eric with the highland cattle he winters out to keep the grass short, helping the choughs' feeding habits. They search for invertebrates in the soil with their sharp beaks. The cowpats are important, too, for they harbour insect life which the young chough, with their shorter beaks, can feed on.

OPPOSITE:
The feature showed how affectionate and cuddly the pups were and how they suckled my skin

162

As filming began at Kildalton, Michaela visited Uisge in the bath who made little 'woo, woo' sounds and rolled on his back for a tickle. Hannah and Fifi sat with Michaela and myself at the courtyard bench-table where we cut up and weighed fish. Michaela helped by carrying Sophie outside and the filming continued with an interview on bringing up seal pups as Michaela helped feed Uisge and Sophie in the courtyard. We took them down to the bay for a swim, the seals following when set down in the grassy field, but Michaela decided the sea was too cold for her to join in. Don brought the Seal Rescue boat round from Cnoc Bay to the pier and George and Michaela went out to the Seal Sanctuary where they filmed the colony where Sophie and Uisge had been born. The cameraman took film of the two sunbathing outside – Uisge on his back, scratching his neck lazily with a flipper and Sophie swimming in the courtyard tank. The feature showed how affectionate and cuddly the pups were with me and how they suckled my skin. Michaela spoke of how I was imitating the close mother-and-pup relationship of common seals that I had observed in the wild. They filmed me playing the violin to the seals and recording in the studio, finishing with the whole family and Michaela dancing crazily on the beach to one of my more lively songs – 'Special Smile'.

An unusual visitor to Islay that summer was the 'Truce Goose'

An unusual visitor to Islay that summer was the 'Truce Goose'. George Wyllie, an internationally renowned sculptor famous for his outsize creations, built an enormous model of a barnacle goose and set it on the far side of a small loch by the main Port Ellen to Bowmore road for all to see. As its name suggests, the sculpture symbolised the agreement which was made between farmers, landowners and conservationists after years of contention over the thousands of wild geese that spend the winter in Islay. The island plays host to about a third of the world's population of the rare Greenland white-fronted goose and two-thirds of the population of the barnacle goose. Although protection is meant to be guaranteed under various laws and international agreements, they have been slaughtered in some areas under the excuse of crop protection. In an effort to offer more positive protection, the Scottish National Heritage made further agreements with the owners of the most important sites, creating sanctuaries for the geese. In addition, the RSPB bought Gruinart Flats as a reserve.

Outside these areas the farmers were paid to scare the geese off their precious grass, aided by people employed in special government schemes. But the scaring was not working to the farmers' satisfaction, for many geese persistently returned to areas outside the sanctuaries. Farmers felt it would be better to welcome these feathered guests on to their grazings and have the government pay for their bed and breakfast. In the winter of 1992/93 the government implemented a goose-management scheme in which farmers are paid £9 per goose on their land, based on the average number recorded out of several counts. The geese seldom land on Kildalton but, on hearing of the scheme, my ever-hopeful husband sent me out into the fields with the violin in case, like the seals, they were attracted to my music. It turned out to be a wild-goose chase.

The erection of the 'Truce Goose' celebrated this new accord reached with the farming community. George Wyllie rowed Hannah across the water in a small inflatable to see his creation close-up. To her, it was like something out of *Alice in Wonderland*.

We had another visit that summer from Jim Vollmar who, after his experience with Aqua, had no qualms about sharing a bedroom with Sophie. In the article he wrote for the *Vegetarian Magazine*, he described how our family would go out in the evenings to the public road and rescue toads that would otherwise be run over by the traffic. Their numbers are declining in Britain – thought to be caused by rising pollution. The children delighted in this rescue exercise and sometimes we would keep toads of all sizes in a box with grass bedding until releasing them somewhere safe the next evening. One day, Hannah thought she would try playing the violin to a

There was a stark contrast between the little seal and the diver with all the equipment he needed to do what she did naturally in her world

large toad and sat it on a rock in front of her. She was disappointed that her music hadn't made the toad turn into a prince, but we explained that it was probably only because it was not the same as a frog.

After Sophie and Uisge had been filmed for the *Really Wild Show*, the word got round the BBC Natural History Unit and we were asked if the two pups could be used for a documentary series called *Living Dangerously*. One of the programmes, 'Have Fish had their Chips?', was to be about fish stocks in relation to the animals of the sea and man's fishing practices. Having heard that the seal pups went out into the sea with me, they were hoping to film underwater shots of seals in a more controlled way than would be possible by going out amongst a wild colony. They wanted the seals to be flown down to Cornwall. We agreed that they could take part as long as I would be present at all times – and the television crew would have to come to Islay.

One day Hannah thought she would try playing the violin to a large toad and sat it on a rock in front of her

Laurie Emberson, a trained animal handler, and the cameraman arrived on a private plane with the film director and all their diving and filming equipment. I didn't know how the seals would react to a complete stranger in this situation, but Laurie was gentle and patient and gradually gained their confidence, meeting them first in the living-room then going for a swim with them with his dry-suit on. Once the director was satisfied all was well, he left the island. The filming took several days and for much of the time we went to the sea by Port Ellen lighthouse as the deep water was clear there.

Here Laurie took Sophie exploring and the cameraman took film of her discovering a lobster by a creel, swimming over clams on the seabed and following a flounder through the water. There was a stark contrast between the little seal and the diver with all the equipment he needed to do what she did naturally in her world. When they finished each shot they would come to the surface and I would call Sophie back to the shore. We also filmed in the next bay along from the one below the farmhouse, reaching it by boat. Beautiful shots were taken of Sophie and Uisge playing together underwater amongst the seaweed. Then the film crew wanted to film Sophie hauling out in a particular spot and I had to crouch behind a rock out of the camera's view as I called her. She was a little star and was used for

As they waited at the living-room door, like two dogs, to be let in

most of the filming because Uisge got over-excited and swam too fast for the camera.

Both seals were plump and strong now and, seeing Sophie lying in the seaweed during a break in filming, I realised that the time was fast approaching when they would leave me. Even with the slide, I was finding it difficult taking them in and out of the large tank and, as they grew older, they became more independent and aggressive.

But I had two beautiful films to remember them by. I was so proud of the two pups when I saw them in the *Really Wild Show* and *Living Dangerously*, in which they had been used to raise awareness of the ecology of the seas. One of my lasting memories of them is as they waited, like two dogs, at the living-room door to be let in, before flopping in and settling down to watch *The Little Mermaid* on television.

TEN

Iona

In the weeks that followed after Sophie and Uisge left us, I would scan the bays of the peninsula and search among the seals gathered on little islands close inshore. Sometimes I took the opportunity to go out in the Orkney longliner with other members of the family, hoping to recognise one of the 'twins' amidst the seal colony or swimming in the water. Neither put in an

Sometimes I took the opportunity to go out in the Orkney longliner with other members of the family

appearance, however, and although I knew this didn't mean they were not all right, I would have loved to have seen them and had my mind put at rest.

Then one sunny day, like an answered prayer, I found Uisge hauled out on a rock on the little island below our house. I had to resist the urge to go down to the pier and call, and watched him instead through the binoculars. He was on his own that day, but he returned often to the same rock, sometimes with other seals joining him nearby. But there were no sightings of Sophie and I feared something had happened to her.

Several more weeks were to pass before I began receiving reports of two seals playing together in the bay below the Fairy Hill. I was so delighted when it turned out to be Sophie and Uisge, and sometimes Aqua was with them. By now I had seen a lot of Aqua and, although she was wild, she spent much of her time in the environs of Cnoc Bay and would come closer in than the other seals if she heard us by the shore.

In early spring, every time the weather permitted, we spent many hours working in the woodland by the sea behind the castle. Taking food to cook on a small wood fire on the pebbled beach, I would then build the bonfire up, burning the large rhododendron branches that George cut as he cleared the old paths. Hannah and Fifi played happily on the beach and among the trees.

One day Aqua bobbed up with a large pink-coloured fish in her mouth. I had never seen this with any seal before; thrilled, I called out, 'Clever girl, Aqua.' We watched with astonishment as a seagull swooped down and cheekily grabbed the fish right out of her mouth and flew away! Aqua had attempted to avoid the bird by ducking down beneath the waves. She rose again after a few seconds and looked around in bewilderment. On another occasion we were working close to the dùn and Aqua had been seen popping up and gazing our way now and then from afar. Suddenly, she surprised us by appearing only a few yards offshore and, reminiscent of the silly, playful moods we knew and loved her for, she turned and broke into a whole series of stunning porpoise leaps across the little bay towards the dùn. We ran the few yards round to view the next bay and were amazed to see her still continuing her acrobatics until she was out of sight past the next promontory.

As the days lengthened and warmed, wild hyacinth embraced the woodland, purple rhododendron buds swelled, and here and there beautiful fungi hugged the tree bark. The children looked like fairies as they sat amid a lavish haze of mauve and the surroundings gave a perfect setting for their games involving the 'little people'. I told them my own stories about the seven gnomes who cared for the environment.

The children looked like fairies as they sat amid a lavish haze of mauve

THE SEVEN GNOMES

See the world,
the beauty of long ago,
still survives where clear, pure waters flow.
In the wild,
where man has yet to go,
free the land where life can live and grow.
The Seven Gnomes have awakened now,
and you can help them too,
the friendly gnomes,
they have returned,
to make the earth as good as new.
Listen to the gnomes as they work and play,
you might see one today.

A hillock freshly grown with shoots of fern and bracken became their own fairy hill where the friendly sprites left them sweets half-hidden in the grass. Some 70 years earlier Talbot Clifton's daughter, Easter, had found the wee folk equally responsive with letters left under the trees. Down on the shore, sea-slaters crawled busily and a host of hoppers jumped wildly in the seaweed. The children turned stones and chose treasures among the shells while the wild mute swans called in for titbits with fluffy cygnets in tow.

As the seals have their legends, so do the swans, for in folklore they are known as the enchanted children of kings. One story tells of the beautiful swan princess of Islay who changed to human form and married a fisherman. They had three children but the swan wife had to leave him after he broke his promise to leave half of his catch at the shore for the wild swans, her brothers. Happily, though, she was able to return after seven years when she took her husband and their daughter to Tìr nan Òg – the land of perpetual youth.

One summer, before I had my own children to play with on the beach, some musician friends of ours came on holiday from London, and I was touched by the fascination their eight-year-old had shown in the bounty of nature found on these Hebridean shores.

DANIEL

Daniel likes to play on the shore,
when the summer sun does shine,
finding shells and pebbles there,
to while away the time.
Each shell a memory;
bring your memories to me,
oh, Daniel, let me share your summer wine.
There'll always be a place in my heart
for the boy with the golden hair,
every wonder nature brings,
a treasure, oh so rare,
brings out the child in me,
with each new discovery,
oh, Daniel, what a world we have to share.

And, Daniel, you will grow from a child
and become a fine young man,
and you'll find that things don't always
work out as we plan.
But life holds many joys,

and men must grow from little boys
who used to love to play upon the sand.
Who used to love to play upon the sand.

But life is full of joys,
and men must grow from little boys
who used to love to play upon the sand.
Who used to love to play upon the sand.

In June 1994 I was five months pregnant, and with the arrival of the pupping season, I was apprehensive of finding a seal in need, not knowing whether I would be able to cope. I was soon to be given the opportunity to find out, though. We were out in the boat on a hot afternoon and there she was, abandoned and waiting to die. Without her mother's milk she had become very dehydrated and, if another day had passed before we arrived

We named her after this beautiful island – Bride

173

Bride was happy to be trundled about in the wheelbarrow

on the scene, it would probably have been too late. Thankfully, there was nothing else wrong with her and she soon recovered.

I had picked her up from a rock on the shore of Eilean Bhride, and we named her after this beautiful island. Bride, pronounced Breej, can be translated to Bridget, or rock elf. I had often played the violin to the seals from that very boulder, as it was flat and comfortable. Bride was an absolute delight to care for as, for the first time – apart from her very early days – I was raising a perfectly healthy seal. This very fact, though, nearly spelt disaster.

She had been in the spare-room bath for less than a week and I only had a small barrier up as she was so young, and I was judging her strength as compared with all the previous pups'. A couple of yards away from the bath I had placed a heat lamp by the door connecting the bathroom to the bedroom. Bride had a slightly low temperature and I put the lamp on at night. It was set upright in a box with a central hole I had made to keep the lamp steady. Early one morning I was doing my rounds attending to the animals. We had a friend, Allen Bryson, sharing the room with Bride, but he had gone out to fetch George from the airport. I was going to go up the field to bring the ponies down but I thank God that for some reason I decided to feed Bride first.

Stepping in from the corridor I found the room full of acrid smoke and, on rushing to the bathroom, I found Bride lying on the floor and the connecting door ablaze. She must have been the first seal ever to have started a fire. I hurriedly picked her up and rushed her out to the courtyard tank, returning to the fire which I put out with wet towels. Bride had hauled herself over the barrier and knocked over and broken the lamp which set fire to the carpet and the door. Poor Allen was shocked when he returned to his bedroom, and especially so as the reason he was staying with us was that on the previous Guy Fawkes' Day he had had a lucky escape when a residence on the estate had suffered a serious fire. The incident made me realise how sick the other pups I looked after had been when they took so much longer to regain their strength.

When teaching Bride to feed I was extremely careful not to receive any cuts on my fingers and wore thick protective gloves – even though they make it harder to feel what you are doing. I could not risk any infection because I knew that the antibiotic effective in dealing with seal finger would be liable to discolour my developing baby's teeth. My stepson, George Junior, took on the heavier handling tasks as Bride grew sleek and fat, going through all the feed-learning stages like a dream. After three days she even opened her mouth for the plastic tube. There is no doubt in my mind that the fact that I was pregnant somehow attracted her to my breasts for, when she was very young,

Going in and out of the doors that lead from the courtyard to the living-room as she pleased

Dusky and Bride would meet in the courtyard and tentatively touch noses

It was as though she regarded herself as one of the children

she always nuzzled me in that area.

I, of course, was growing steadily bigger and had to be careful not to strain myself, but Bride was happy to be trundled about in the wheelbarrow, and I wondered if she was the first seal ever to ride in one. She was friendly with all the family, following Hannah and Fifi when flopping around on the castle lawn and climbing into George's lap as he sat in the living-room. She was obviously quite at home, going in and out of the doors that lead from the court-yard to the living-room as she pleased. Hannah was given a black female kitten for

She would wear her *Little Mermaid* T-shirt as she played with Bride

her birthday. Dusky and Bride would meet in the courtyard and tentatively touch noses.

When Bride had been with us for three weeks, she was introduced to the large outside tank. She made lonely 'wooing' cries for a little while but soon settled down and astounded us with the games she played with balls and other children's toys. She would splash up and down wildly, twisting and turning as she swam with a red tennis-sized ball in her mouth, and with her nose she bounced a plastic football repeatedly against the side of the tank. I would be woken early in the morning to the sound of her energetic capers.

When taken to the bay, tidal waves were produced by her amazing bursts of speed, but then she would go for a quiet wallow amongst the thick

Bride seemed reluctant to abandon the shallow sides of the bay

seaweed that edged the shore. Through deep blues and verdant greens she glided, gently stirring the tranquil underwater garden beneath her. The children loved their seal playmate and Bride was so placid that, with the water's support, they could hold her safely. They would laugh at the funny expressions she put on her face as she anticipated her picnic of fresh, filleted herring on the beach. Sometimes she seemed to be waiting for the family to join her in the water. It was as though she regarded herself as one of the children. Hannah loves the Disney stories and she would wear her *Little*

Mermaid T-shirt as she played with Bride. I, too, enjoyed cuddling her in the water – where all the seal pups were at their most relaxed and Bride especially liked lying on her back in my arms as I tickled her on the tummy.

All too soon, it was time for her to leave. I was losing one baby and looking forward to the birth of my own. Bride seemed reluctant to abandon the shallow sides of the bay. Then Aqua appeared and with wonder we watched as she guided Bride out to sea.

I flew to the mainland to await the arrival of my baby in Paisley Maternity Hospital. Almost five years earlier I had given birth to Fifi in the same hospital and, as Hannah had never been separated from me before then, I was anxious for her wellbeing. George agreed to look after her at a nearby hotel with my stepdaughter, Sarah. He assured me everything was fine.

On arriving at the airport on the way home with the baby, we walked through the shopping area and George tried to usher me past the newspaper stand. I soon discovered why when I was shocked to see a large photograph of my husband and Hannah staring at me from the front page of the *Daily Record*. The headline banner read 'Killer Toys from Santa'. George then told me how, when visiting Santa in his grotto at a Glasgow store, Hannah (then two and a half years old), had been given a potentially deadly set of jewellery in a bag. On returning to the hotel, Hannah had gone into a corner while George and Sarah unpacked the shopping, and stuffed her mouth full of the brightly coloured small parts. She nearly choked to death – only being saved because George and Sarah heard her. The *Daily Record* had taken up the story, using it to start a safety campaign for children at Christmas.

But that had been five years earlier, and this time, having given birth to Iona, I relaxed with my gorgeous new baby in the hospital for a few days. George returned to Islay and was looking after Hannah and Fifi at home. Surely nothing would happen this time. John Robins of Animal Concern visited me and during our conversation he asked me what I would do if a seal turned up on Islay needing attention. I laughed about it and assured John it was most unlikely to happen in November.

However, the evening before I was due to fly home, a nurse came to my bed and said there was a phone call for me. I followed her to a small private room where I took the call. It was Don Bowness. I thought, 'How nice, he's rung to congratulate me on the baby.' This he did, but then said, 'We've got another baby for you here.' Don wanted instructions on how to attend to a grey seal up until my return the next day. As I went through the necessary information with him, I was thinking all the time that this was a belated April Fool's joke. But when I rang George he told me that he was

I was shocked to see a large photograph of my husband with Hannah staring at me from the front page of the *Daily Record*

just hurrying over to Don's house with the seal-feeding equipment, and I realised it was no joke. I would have to operate under the strictest hygiene conditions with my newborn baby to look after as well.

We named him Sandy after the girl who had found him while walking with her husband on a beach on the Oa peninsula to the southern side of Port Ellen. He was past the weaning stage, having lost his white fur, but was pitifully thin and malnourished. He was also exhausted and for many days did nothing but sleep deeply between feeds. I attended to him with full protective clothing, including a mask to prevent my catching any airborne germs. He went from strength to strength and, after about a week, was eating whole fish by himself. Local fishermen were especially generous with gifts of whiting to help with his food supply, as fresh herring was extremely difficult to get at that time of year.

Sandy was not a gentle soul, and when I was tube-feeding in the first few days, I felt as though I was getting into the bath with Jaws. Sitting astride him one evening, the lights went out in a power cut in mid-feed. After a few seconds of rising panic, I decided to picture everything I would normally do in my mind and continue by touch in the pitch blackness. I had gone through the tube-feeding procedure so many times before that I surprised myself, and I managed to complete the job before gingerly stepping out of the bath and groping my way back to the family.

Once Sandy got more used to me, I was able to carry him outside in the courtyard. I posed for the family album with my two newborn babies. I kept Sandy at a safe distance from the children for he certainly was not as cuddly as the common seal pups. I was glad to give him a chance of life though, and recalled with sadness how another grey seal pup, still covered in soft, white fur, had been so far gone when he was picked up that he died in my arms on the way to the vet's surgery.

It was hard work looking after a seal so soon after giving birth to Iona, and I was relieved when the RSPCA Wildlife Hospital in Norfolk offered to take him for me. Dougie Walker, the veterinary nurse who had chaperoned Morag on her travels, was coming up to Islay for a short holiday and arrangements were made for him to collect the seal. Meanwhile, the fishmonger was still searching up and down the country and, eventually, located what we were looking for. Santa put in an appearance and gave Sandy a Christmas present of fresh herring which had come all the way from Iceland.

To give him some exercise prior to his long journey, I took Sandy into the woods by the castle where he lolloped around in the damp autumn leaves. George and I were amazed when, without a glimpse of the sea, Sandy took off determinedly in the direction of the pier. I soon caught up with him and, gathering him up in a large towel, took him back to the farmhouse where his new minders awaited with a special animal travelling-box.

As the winter wore on, fierce gales and high tides brought the usual unsightly assortment of discarded rubbish on to the beaches. When riding Jason round the shore, I had developed a habit of dismounting and picking up any fishing-net, securing large pieces in clefts of rock above the beach for collection in the spring.

The danger to wildlife of netting had been brought home to me by an incident which had occurred some years before in which a fallow buck, with fishing-net caught in his antlers, had become entangled in a fence. If I had not found him, he would certainly have slowly starved to death. Of course, the deer was terrified when George and I approached, and was in danger of breaking his neck or injuring his rescuers as he thrashed about. Fortunately, the vet was already on her way as I had called her out to attend to a sick dog. The deer was tranquillised and the netting carefully cut away. Still slightly dopey from the drug, it took him a moment to realise he had been set free. The buck stood up, looked at us for a second, then bounded off into the trees.

I posed for the family album with my two newborn babies

Santa put in an appearance and gave Sandy a Christmas present of fresh herring

The event was reported in a full-page *TV Quick* article the same week that the Islay *Highway* programme was shown. Highlighting the danger to wildlife of discarded materials washed up on our shores, it asked for the co-operation of readers in removing such rubbish from beaches. Seals and birds become entangled in netting and die, and many seabirds, turtles and whales are known to consume plastic to fatal effect. One study by the American environmental groups estimates that plastic in the sea may kill two million seabirds and

He lolloped around in the damp autumn leaves

one hundred thousand marine mammals each year. A recently published survey by the Tidy Britain Group shows marine litter rising by 40 per cent on Scottish beaches over the last decade. Although there is legislation to deal with this pollution, the problem largely lies in enforcement.

In another incident a fallow buck was found in trouble near the castle. One antler was trapped in the loop of a long length of fishing-rope and the other end was caught in a large tree. Again the deer was thrashing about wildly, but this time he was able to run about. He could kick out dangerously or injure us with his lethal antlers, and the problem appeared insolvable. Even if George could cut the rope at the tree, the deer would have run off with a length still stuck on his antler and eventually would have become caught somewhere else.

As we stood back discussing possible solutions in hushed tones, George had an idea. If he could coax the deer to walk round the tree the rope would tether it tighter and tighter. Slowly and with great caution George

One antler was trapped in the loop of a length of
fishing-rope (Fiona)

He continued in circles after the buck, and the rope
grew shorter (Fiona)

The deer's head was tight against the tree (Fiona)

approached and the deer retreated. He continued in circles after the buck and the rope grew shorter until the deer's head was tight against the tree. The poor frightened animal was now unable to move and it was possible for George to cut the rope next to the antler. The buck ran back into the forest, his antlers now completely free of rope.

We were extremely relieved at the successful conclusion of the rescue and pleased that we had been able to save one of the special castle bucks we know so well. Surprising king buck in a leafy glade, he lifts his handsome face in a regal stare and saunters off into the deep woodland. On still autumn nights we have watched the bucks duel on the castle lawn, listening to the clash of antler upon antler as these princes of the forest tussle during the rut. It was a sad day when I found one of these majestic beasts who had somehow succumbed to the rigours of battle. It is rare, however, for these mating rituals to end fatally.

Sandy had been gone some weeks and I was spending the long winter evenings writing. One night we were in the middle of yet another long power cut during the worst storm of the winter. The powerful gusts of wind roared in from the open sea and shook the house. Pausing in my writing by candlelight, I was chatting to George in the soft glow about the chapter on Morag. I was literally just saying to him, 'Remember how I was missing Aqua and then the phone rang?' when the phone did ring. I could hardly believe it when a woman calling from Port Ellen said that a seal was lying on the grass that edges the village bay.

Leaving Iona in the care of George junior, we grabbed an old blanket, donned layers of warm clothing and headed off in the car as the storm raged around us. We had to pull clear a large pine branch that had fallen across the drive, and the road at Seal Bay was covered in seaweed from the high tide. As we drove round the bay at Port Ellen, big, menacing waves were being driven into the shore by the south-easterly gale. The seal was lying on its side on the shore opposite Councillor McKerrel's house, as though it had come to ask him for assistance. In the car headlights I could see it was a six-month-old common and decided I could handle it by myself. Approaching from the sea edge, I threw the blanket right over the seal in a rugby tackle and, before it had a chance to escape, picked it up and took it to the car. It was a female and I held her in the blanket in the back seat all the way home, organising her bath by torchlight on arrival. She had several small injuries and her breathing was laboured. We called her Gale.

She was badly infested with worms and it was several weeks before she was keen to eat. Because she was ill, she had been unable to cope with the bad storm, being forced to haul out wherever she could. During Iona's naps,

The Fairy Queen slumbers beneath the hill that is blanketed with snow like an iced fairy cake

George would help me feed Gale to keep her strength up, for he had to hand me the fish while I kept her mouth open. Her breathing improved and she became a different seal, waiting at the barrier to be fed when she heard me coming.

As I write by the brightly burning wood fire with Iona tucked under my arm and Gale sleeping peacefully in the bath, we await the imminent return of Sandy from Norfolk. Thinking of his future, I imagine what that little pup could look like in a few years' time. As a male grey seal, the largest mammal in Britain, he could eventually grow to 500lb. I watch the living flames in their perpetual dance as they spirit up the dark chimney, and look back to my childhood in Seal where my love of the violin was born. Standing barefoot on the Palladium stage, I play gentle music evocative of the ethereal ocean mists and the mystic kingdom of the people of the sea. I recall my journey to Islay on the old *Viscount* and my first enchanted

OPPOSITE:
From that otherworld of eel and crab, scallop bed and seaweed jungle, they gaze and dive and return again to contemplate my form

encounter playing to wild seals. I think of all the wonderful moments I have had with seals in my care, and long for Hannah, Fifi and Iona to grow up playing and singing to seals from the shore.

> Today the seals, tomorrow who'll suffer while mankind plays the
> fool?

I hope that our futures and those of the people of the sea will be happier than the sad words of my song suggest.

Outside in the chill winter night, the Fairy Queen slumbers beneath a hill blanketed with snow like an iced fairy cake, and in the still, white forest even the castle shutters are silent.

When I go down to the shore and call across the calm water, my friends appear, gliding effortlessly before their V-wakes. From that otherworld of eel and crab, scallop bed and seaweed jungle they gaze and dive and return again to contemplate my form; and some with soft, familiar faces join me in a keen and singular remembrance.

> People of the sea,
> when I heard your cry to me,
> and I knew that many'd never understand.
> Well I said I'd fight for you,
> with the ones whose hearts are true,
> with those who love the people of the sea.

Postscript

In October 1980 I engaged in a televised debate with a representative of the Orcadian seal cullers. Despite a total lack of scientific evidence to justify their claim that seal numbers adversely affected commercial fish catches, the pro-culling lobby fiercely defended the annual carnage during which thousands of day-old grey-seal pups were slaughtered. However, a combination of activists standing between the rifles and the pups, public outrage at television pictures of seas turned red with innocent blood, and a total collapse of the market in sealskins for the fur trade, soon brought about the end of this mass destruction of the people of the sea.

Campaigners downed a few ales, celebrated this victory and moved on to the next issue. While attention focused on factory farms and vivisection, the seals continued to be persecuted in a secret slaughter carried out far away from the glare of publicity.

In January 1988, I followed up reports of seals being shot on the river Helmsdale in north-east Scotland. Twenty seals had been killed by the Helmsdale River Board who claimed they were protecting salmon stocks entering the river to spawn. Locals, who helped me find some of the carcasses of the shot seals, also informed me that as many as 200 seals were killed every year at the nearby commercial salmon-netting stations.

The publicity which we generated around this case resulted in other reports of seal shootings. On the Mull of Kintyre, a salmon-netsman had wiped out a whole colony of seals. Over on the island of Raasay, I photographed several carcasses of seals which had been shot, washed ashore opposite a salmon-farm. At St Cyrus National Nature Reserve near Montrose, a salmon-netsman who admitted to shooting seals was found to be an Honorary Warden for the Nature Conservancy Council.

The one thing which these and other cases had in common was that no laws were being broken. As long as the correct calibre of bullet is used, anyone with a relevant firearms licence variation can shoot seals which come

189

(Chris Davies)

near any fishing or fish-farming equipment at any time of the year. No proper records are kept of people who shoot seals, or of the number of seals shot each year. With over 300 fish-farm sites, over 100 netting stations and countless lobster-fishing operations, I estimate that, at the very least, 5,000 seals are shot every year around Scotland. One salmon-netsman openly admitted to killing over 90 seals a year at just one site in Morayshire.

In 1988, a farm-worker on Skye wrote to me claiming that seals were regularly being shot at the intensive salmon-unit where he worked. I went to Skye to investigate the claim and was horrified to realise that a salmon-farm owner was shooting seals with a shotgun. Farm-worker Simon Upton bravely agreed to give evidence in court and we were able to secure the first successful prosecution under the Conservation of Seals Act 1970. The guilty party was fined the grand total of two hundred pounds. Two hundred pounds is about the value of 15 out of the several thousand salmon in just one of the floating cages at the farm.

While I was investigating the terrible toll humans inflict on the seal population, a natural disaster wreaked havoc on these animals. During the distemper virus epidemic of 1988, I received a telephone call from a softly spoken young woman from Islay. She told me how she sang to the seals at Kildalton and wanted to do something to help protect these animals. I booked a seat on the Loganair flight to Islay, half-expecting to meet someone who was overdoing it on the very nippy sweeties from the local distilleries.

Less than an hour after leaving Glasgow, I was sitting on a rocky islet, 20 feet behind Fiona as she played her violin. First one dark head broke the surface of the water, then another and another, and yet another. In ten minutes at least 25 seals had gathered to listen to the music. Heads held high above the sea, they drifted past on the tide, dived and swam back to their starting point to float past once again.

These totally wild animals showed complete trust in Fiona. She was part of their environment and was accepted by them as being of no threat. That first visit to Kildalton was a very new experience for me. My usual contact with seals was after human interference had left their lifeless, bloody carcasses lying on rocks or on beaches all over our coastline.

Through Fiona's unique bond with the seals of Kildalton, I was able to raise public awareness of the very real threat humans pose to these magnificent and intelligent creatures. Instead of peddling the horrors of seal shootings, I had a new and far more attractive image to promote to the media. Television, radio and newspaper reporters queued to visit Islay and see Fiona sing to the seals. By controlling these visits and ensuring that the presence of outsiders did not disturb the colony, it was possible to open the eyes of a much wider public to the beauty of seals and the ugliness of how humans persecute them.

Over the last seven years there have been improvements in how seals are treated. Many salmon-farmers, but by no means all of them, have responded to consumer concern and have developed non-lethal ways of deterring seals from their farms. Sonic scarers, properly situated and maintained, security nets and even glass-fibre models of killer whales are now used to frighten seals away from some salmon-farms. However, many fish-farmers, salmon-netsmen, angling bodies and commercial fishermen still regard seals as vermin and kill as many of these animals as they can. Thousands of seals are still shot every year. Seals have been fed fish booby-trapped with explosives and razor blades. Ignoring the fact that human overfishing threatens the very existence of our marine ecosystem, there are growing calls for a mass cull of as many as 50,000 seals. Seals are still a scapegoat of human greed and exploitation, and the fight goes on to achieve legislation which truly protects all seals from destruction by humans.

As well as deliberate acts of violence against seals, there are other hazards of human interaction with these animals. Seals frequently become entangled in lost fishing gear and other flotsam. Modern nylon and plastic ropes can cut these animals to the bone, causing a lingering and painful death. Young seals, left by their mothers for short periods, can be endangered if humans touch them, as our scent can cause the mother to abandon her pup.

If you find a seal which is obviously injured or trapped, it is best to get expert help as soon as possible. Should you come across a young seal which is neither injured nor at risk, then leave it well alone and return several hours later. If it is still there, it is time to call in the experts. Remember, seals, even little fluffy pups, are wild animals which pack a powerful bite and can harbour diseases dangerous to humans. There are various seal sanctuaries and rescue centres around the United Kingdom. Animal Concern, on 0141-334-6014, can give you details of the one nearest you.

I have been privileged to have been able to share a few hours with Fiona and her seals. Through her story I hope you have felt the very special relationship she has with her people of the sea. I hope you will join Fiona in her quest to gain true protection for seals and the fragile environment which they inhabit. All life on earth is threatened by human exploitation and destruction of the marine ecosystem.

John F. Robins,
Organising Secretary,
Animal Concern
September 1995

For more information on the campaign to protect seals and details of how to obtain the picture CD or cassette of Fiona's album *Today the Seals*, contact:

Animal Concern,
Freepost,
Glasgow G3 8BR